CW00972371

The Shuttle Story

The Shuttle Story

John Christopher

Also in this series:

The Concorde Story
The Spitfire Story
The Vulcan Story
The Red Arrows Story
The Harrier Story
The Dam Busters Story
The Hurricane Story
The Lifeboat Story
The Tornado Story
The Hercules Story

Published in the United Kingdom in 2010 by
The History Press
The Mill · Brimscombe Port · Stroud · Gloucestershire · GL5 2QG

Copyright © John Christopher, 2010

All rights reserved. No part of this publication may be reproduced, stored in a retrieval system, or transmitted, in any form, or by any means, electronic, mechanical, photocopying, recording or otherwise, without the prior permission of the publisher and copyright holder.

John Christopher has asserted their moral right to be identified as the author of this work.

British Library Cataloguing in Publication Data
A catalogue record for this book is available from the British Library.

Hardback ISBN 978 0 7524 5174 9

Typesetting and origination by The History Press
Printed in Italy by L.E.G.O. S.p.A.

Half title page: *STS-110 mission specialist Rex Walheim working on the International Space Station (ISS) Destiny laboratory, April 2002.*
Half title page verso: *Two of* Discovery's *STS-124 astronauts work on the International Space Station (ISS), June 2008.*
Title page: *Against a backdrop of clouds,* Endeavour *STS-126 during rendezvous and docking with the ISS, November 2008.*

➤ Discovery *STS-119 riding the giant crawler to launch pad 39A.*

CONTENTS

Acknowledgements	6
Introduction: Origins of the Spaceplane	7
STS – Space Transportation System	14
The Space Shuttle Fleet	21
Countdown to Launch	30
Lift-off!	44
Challenger – A Major Malfunction	52
Living in Orbit	62
Working in Space	71
Spacelab and Spacehab	82
Hubble Trouble	86
Space Stations	94
Returning to Earth	102
The Reusable Spacecraft	109
End of an Era	115
Appendix 1 – Seeing the Space Shuttle	119
Appendix 2 – Technical Specifications	122
Appendix 3 – Glossary	127

ACKNOWLEDGEMENTS

Almost all of the images in *The Shuttle Story* are from the archives of NASA, and any from other sources are acknowledged individually. For her continued support and proof-reading diligence I am also indebted to my wife Ute, and for Anna and Jay. I hope this account of the Space Shuttle will serve to convey some of the magic and excitement of the continuing exploration of space.

John Christopher

The quotations in this book come mainly from official NASA sources and in addition, a number of other publications were consulted including: *Space Shuttle Launch System 1972–2004* by Mark Lardas, *Heroes in Space – From Gagarin to Challenger* and *Zero G* by Peter Bond, *Liftoff* by Michael Collins, *To Rise From the Earth* by Wayne Lee, *Challenger: A Major Malfunction* by Malcolm McConnell, *Entering Space* by Joseph P. Allen, *Living in Space* by Peter Smolders, *The Space Telescope* by David Ghitelman and *Disasters and Accidents in Space* by David J. Skayler.

INTRODUCTION: ORIGINS OF THE SPACEPLANE

In 1969 the Apollo 11 astronauts journeyed to the Moon atop the colossal Saturn V rocket and a wave of patriotic public support. President Kennedy had vowed to put an American on the Moon before the decade was out and NASA had pulled out all the stops to fulfil that momentous dream. It was an incredible and courageous undertaking; the pinnacle of man's achievements in space. Originally a further nine Lunar landings had been planned and NASA had ambitions to establish a base on the Moon and to construct an Earth-orbiting space station from which a manned mission would embark on its way to Mars.

But no sooner had the Apollo 11 astronauts returned to Earth and the bubble burst. Of the intended twenty Apollo missions, three were cancelled in an atmosphere of public indifference and political resistance to the space programme's prohibitive costs. Looking to the post-Apollo era NASA's focus turned away from the exploration of space and back towards our own fragile planet. Facing tight budgetary constraints all talk of going to Mars went out of the window in order to concentrate the remaining resources on the space station.

The HL-10 was one of five lifting-body designs tested at NASA's Dryden Flight Research Center, Edwards, California. It flew on thirty-seven occasions and is shown shortly after landing on 1 January 1969, with pilot Bill Dana taking a moment to admire the Boeing B-52 aircraft which had dropped him.

▶ *This display at the NASM, Washington, shows the evolution of the Shuttle concept from the fully reusable two-tier winged spaceplane (left) to the final design with twin boosters and a disposable external tank.*

to the Earth. Servicing the space station and maintaining America's presence in space would require a revolutionary new type of reusable launch vehicle, a winged 'spaceplane'. This would drastically reduce costs and place the business of putting payloads into orbit – whether astronauts, satellites or space station components – on such a regular and routine footing that it would become just that: a business.

Preliminary design studies had begun in October 1968 to produce a space launch system that did away with the wasteful once-only rocket boosters used up to that time. Of the Saturn V's 363ft (111m) only the very tip of the stack, the capsule, returned to Earth and even that was incapable of being used more than once. A leading design by NASA's chief designer, Maxime Faget, featured a two-

This would provide a permanent platform from which astronauts and scientists would conduct their studies and experiments, with the emphasis on bringing benefits

tier system with a winged booster the size of a Boeing 747 carrying an orbiter up to a an altitude where it could separate and continue on its way. Both craft would re-enter the atmosphere independently, to land horizontally on a runway much like conventional aircraft. Unfortunately, full re-usability brought a hefty price tag, as the booster craft would need to be huge. Instead, a partially reusable design gained favour, with a winged orbiter attached to an external propellant tank and booster system which could be refurbished.

Originally the orbiter was to have jet engines to assist manoeuvring in the atmosphere after re-entry, but these were omitted in the final design to reduce complexity and increase payload capacity. The latter was a significant factor in shaping the orbiter, not only to maximise commercial payloads, but also to satisfy the military. During the 1960s the US Air Force had been working towards several independent projects aimed at maintaining a manned military presence in space. These included the X-20 Dyna-Soar spaceplane and the Manned Orbital Laboratory (MOL), both of which were cancelled. To ease the

▲ 5 January 1972, Dr James D. Fletcher of NASA shows President Richard Nixon a model of the proposed Space Transportation System (STS) on the day that the President announced that the USA should proceed with the programme.

1975 representation of the STS which became better known more simply as the Space Shuttle.

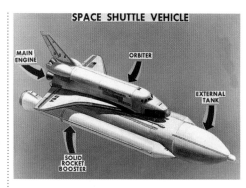

A wind tunnel test, conducted at the John H. Glenn Research Center, to investigate what would happen if the main engines were to fail during launch.

financial burden of developing the Space Shuttle, NASA courted the involvement of the military. Such partnerships were not uncommon, but compromises had to be made. While some facilities would be shared, the Air Force stipulated the ability to launch and land at the Vandenberg Air Force Base (AFB) in California. Another requirement was to launch satellites southwards on a trans-polar orbit, without the usual assistance from the Earth's rotation, and as this would require greater thrust it necessitated a greater payload and hence a bigger Shuttle.

To get the Shuttle into orbit NASA considered various options, including the development of the existing Saturn V lower stage booster, a single large solid fuel booster, or a pair of smaller ones. At the time there was some concern about the reliability of the solid rockets but in the end they were selected in order to keep development costs down.

With the final design agreed, contracts were awarded to several US aerospace companies. Rockwell International (formerly North American Aviation) would build the Space Shuttle's Orbiter, Martin (later to become Martin-Marietta) the external tank, and the contract for the rocket motors went to Thiokol (now Morton-Thiokol). Rockwell was to build two Orbiters initially, OV-101 and OV-102, plus a 'structural test' airframe designated

◁ Enterprise in flight in 1977, shortly after being released from the Boeing 747 Shuttle Carrier Aircraft (SCA) over Rogers Dry Lakebed. This was the second of five free flights to verify the Orbiter's handling characteristics and aerodynamics.

'The USA accepts the basic goal of a balanced manned and unmanned space program. To achieve this goal the United States should develop new systems of technology for space operation … through a program directed initially toward development of a new space transportation capability.'
Presidential Space Task Group, 1969, two months after the Apollo 11 landing on the Moon.

Did you know?
All of the Space Shuttle Orbiters are named after famous sailing ships of exploration; *Columbia, Challenger, Discovery, Atlantis* and *Endeavour*.

Commander John Young and Pilot Robert Crippen were selected as the two-man crew for the first manned launch of STS-1 which took place in April 1981. Young was a veteran of the Gemini and Apollo programs and in 1972 he had walked on the Moon as commander of Apollo 16.

STA-099 (see 'The Space Shuttle Fleet'). Construction of OV-101 began in 1974 and christened *Enterprise*, it began a series of Approach and Landings Test (ALT) flights. Carried aloft by the special Boeing 747 Shuttle Carrier Aircraft (SCA) these culminated with five free flights with *Enterprise* released at altitudes ranging from 19,000 to 26,000ft (5,790–7,925m) to make an unpowered gliding approach for a runway landing.

On 12 April 1981, and several years behind schedule, *Columbia* became the first re-usable manned spacecraft to enter space. It launched from Pad 39-A at the Kennedy Space Centre (KSC) and after thirty-seven orbits, landed at Edwards AFB. In command was John Young, a veteran astronaut from Gemini and Apollo, accompanied by Robert Crippen, a former military astronaut who had transferred to NASA when the MOL programme was

◁◁ *STS-1 mission Commander, John Young, seated in the commander's station on the port side of* Columbia's *forward flight deck.*

◁ *Lift-off for the world's first reusable spacecraft.* Columbia *rises on two columns of flame from launch pad 39-A at the Kennedy Space Centre, 12 April 1981.*

▲ *Mission patch for STS-1, the first orbital flight of the Space Shuttle.*

cancelled. For the initial Research and Development flights, STS-1 to STS-4, the Shuttle flew with only a two-man crew as *Columbia* only had two ejector seats.

Starting with STS-5 the spaceplane was finally ready for business, and NASA embarked on an ambitious launch programme that would span more than thirty years.

STS – SPACE TRANSPORTATION SYSTEM

The Space Transportation System (STS), to give the Shuttle its official name, is the most complicated and expensive flying machine ever built. It consists of four major components: the winged Orbiter; a large external fuel tank for the main engines; and a pair of rocket boosters. The Shuttle launches vertically, just as with a conventional rocket, and standing on the launch pad the STS stack measures in at 184ft (56.14m) in height.

THE ORBITER

The world's first spaceplane, the Space Shuttle's Orbiter carries astronauts and payloads into low Earth orbit and re-enters the atmosphere to land on a runway as a glider. At 122.2ft (37.2m) long with its double delta wings, swept at 81° at the leading edge and 45° at the outer, plus a large vertical stabiliser, it resembles a conventional aircraft. The Orbiter's structure is primarily of aluminium alloy, although the engine structure is mostly of titanium.

➤ *This dramatic view of Atlantis STS-79 on the crawler-transporter clearly shows the Orbiter, the orange-coloured ET and the twin Solid Rocket Boosters (SRBs) to either side.*

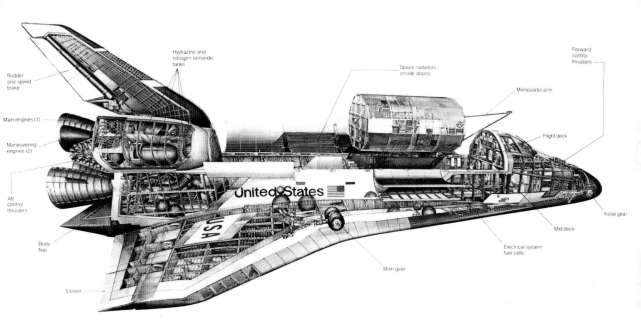

The crew cabin within the nose is arranged on three levels with an upper flight deck, then mid-deck and below that a utility area. In practice it usually operates with a crew of five to seven astronauts, although the largest crew so far has numbered eight.

Occupying the mid section of the fuselage is the large payload bay 60ft long by 15ft wide (18 x 4.6m). This is vented to

▲ *Cutaway of the Space Shuttle Orbiter with the payload doors open and the manipulator arm moving an item of space cargo.*

> *The Orbiter is covered with silica-glass thermal insulation tiles to protect it from the intense heat of re-entry into the atmosphere.*

Did you know?
Only three Apollo astronauts went on to fly the Shuttle. John Young commanded the first Shuttle flight into space with STS-1 in 1981, and Thomas Mattingly commanded STS-4 the following year. Fred Haise took part in the ALT tests with *Enterprise*.

'The dream is alive again.'
John Young, after the flight of STS-1 Columbia

the vacuum of space and the two payload bay doors, which have heat radiators on their inner surfaces, are kept open while in orbit. Inside this bay is the robotic arm known as the Remote Manipulator System (RMS), or Canadarm, used to deploy or retrieve payloads, or to act as a working platform for astronauts and provide a camera arm to inspect the Orbiter's Thermal Protection System (TPS). Thousands of individual silica-glass insulation tiles cover the underbelly, bottom of the wings and other heat-bearing surfaces of the Orbiter, resisting temperatures of up to 704°C by radiating 90 per cent of the absorbed heat back into the atmosphere.

At the rear of the fuselage are the three Space Shuttle Main Engines (SSMEs), each of which can produce 1,668,000 Newtons of thrust, and these are fuelled by the external tank. Their nozzles can be gimbled, or tilted, to provide steerage during the lift-off. A pair of smaller Orbital Manoeuvring System (OMS) thrusters, located either

◀◀ *Space Shuttle Main Engine (SSME) undergoing a full-power firing in 1981 as part of tests to increase the amount of thrust available in order to carry heavier payloads into orbit.*

◀ *Discovery STS-124 touching down at the Shuttle Landing Facility at the Kennedy Space Center (KSC) in June 2008.*

side of the SSMEs, provide thrust for orbital manoeuvres, and a cluster of small altitude-control rockets located in the nose and at the rear allow smaller thrust inputs for finer control of attitude and orientation. The aft area also houses three Auxiliary Power Units (APUs) which provide pressure for the hydraulics to gimbal the main engine nozzles, control elevons and rudder during the descent, and also deploy the landing gear.

EXTERNAL FUEL TANK

As the Orbiter is not large enough to carry its own internal fuel tank for the SSMEs, the fuel is fed to the engines from the large

> *An External Tank (ET) being transferred for shipment by barge from NASA's Michoud Assembly Facility, in New Orleans, to KSC.*

Did you know?
The Space Shuttle was one of the first aircraft to incorporate fly-by-wire technology which does away with mechanical or hydraulic linkages to the control surfaces.

External Tank (ET). Internally the tank is split into two compartments holding the cryogenic liquid rocket fuel, oxygen and liquid hydrogen. As the boiling point of these fuels is around -200°C, the ET is coated with a layer of spray-on foam which prevents ice from forming on the exterior. This foam gives the tank its rusty-orange colour, although on the first few launches the ETs were finished in white. This was quickly abandoned as the extra layer of paint added unnecessary weight. The ET is the only part of the STS which is not reusable and it burns up during re-entry.

SOLID ROCKET BOOSTERS

As the combined mass of the Orbiter and the ET is too great for even the SSMEs to lift, a pair of Solid Rocket Boosters (SRBs) provide over 80 per cent of the

power required. Each SRB produces an incredible 11,790,000 Newtons of thrust during the first two minutes of lift-off. They are constructed in four main segments which stack one upon the other, capped with a nose cone housing the igniter, electronics and parachutes for their safe return. Once they have been ignited the SRBs cannot be turned off or throttled

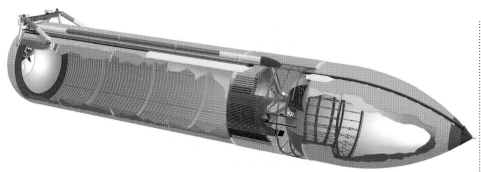

◁ *Cutaway of the External Tank (ET) revealing the upper compartment for liquid oxygen and the lower one for liquid hydrogen. Liquid oxygen is far denser than liquid hydrogen, which is why the upper tank is smaller in volume.*

back. Unlike conventional rockets, which work by combining two liquid propellants, SRBs are filled with a solid fuel which has a consistency like hardened putty.

'The fourth landing of the *Columbia* is the historical equivalent of the driving of the golden spike which completed the first transcontinental railroad. It marks our entrance into a new era.'

President Ronald Reagan – on the final test flight of the Space Shuttle, 4 July 1982.

This is done in individual SRB sections to ensure a uniform mix of the chemicals and to facilitate easier transportation as each complete booster weighs 1,300,000lb (590,000kg). Final mating of the sections takes place within the Vehicle Assembly Building (VAB) at Kennedy Space Centre.

The Shuttle is designed to reach relatively low orbits ranging between 115 and 400 miles (185–643km), and it is not capable of taking astronauts deeper into space, or

Did you know?

Until an update in 2007, Shuttle missions could not be allowed to straddle the year-end roll-over from December to January because the software required resetting.

> *Twin pillars of raw power from the Solid Rocket Boosters (SRBs) thrust Atlantis STS-43 on its way to orbit, August 1991.*

to the Moon for example. Most missions last anywhere from five to sixteen days, and the longest single mission was STS-80 which spent 17.5 days in orbit.

The Shuttle programme has not been without its critics, especially when it comes to the cost, which is estimated at an average of US$1.5 billion per mission. In some respects the seeds of this criticism lie in NASA's inability to meet the over-blown promises it made when originally selling the programme to the politicians. Even so, since 1981 the STS has launched more than 3,000,000lb (1,360,000kg) of cargo into orbit and more than 600 crew members. Considerable improvements have been made, in particular with regards to the payload carried, and NASA claims to have trimmed the operating costs by as much as 40 per cent in recent years.

THE SPACE SHUTTLE FLEET

In order to fulfil its goals and ambitious launch schedule, NASA would require not just one Space Shuttle Orbiter, but a whole fleet of them. In total six were built, and although they may look like peas in a pod at first glance, a continual process of improvement and upgrades – partly in response to the catastrophic loss of two of its craft – has resulted in a range of variations in these incredible flying machines.

ENTERPRISE OV-101

Construction of the first Orbiter began in 1974. Originally, it was to be called *Constitution*, until a concerted write-in campaign by *Star Trek* fans persuaded NASA to re-name it *Enterprise*. Given the Orbiter Vehicle Designation OV-101, it was built to conduct test flights within the atmosphere with a view to conversion to orbital status after *Columbia*. By then, however, many design details had changed – particularly with fuselage and wing weights – and the upgrade was deemed too expensive. *Enterprise* had a smaller tail than later Orbiters and a number of subsystems, including the main engines, were not fitted. Initiating the ALT programme, it first flew on 18 February 1977 perched atop a Boeing 747 Shuttle Carrier Aircraft (SCA). Eight unmanned and manned flights were conducted in this fashion followed by five free flights in which *Enterprise* separated

> 'I know how to never have another Challenger. I know how to never have another leak, and never to screw up another mirror, and that is to stop and build some shopping centres in the desert.'
>
> J.R. Thompson, NASA deputy administrator.

Did you know?
The *Enterprise*, named after a massive write-in campaign by *Star Trek* fans, was intended to be refitted for orbital flight. Unfortunately it was passed by in favour of a new Shuttle, and *Enterprise* never did head off into that final frontier.

to land as a glider. *Enterprise* was involved in further ground tests including mating with the ET and SRBs on the launch pad in 1979. In semi-retirement it went on tour riding piggy-back on the SCA, and in 1985 became the property of the Smithsonian Institution. Following the *Challenger* disaster NASA briefly considered refitting *Enterprise* for active service but decided to construct *Endeavour* instead. OV-101 is now a museum exhibit. (See Appendix 1 – Seeing the Space Shuttle.)

➤ *Columbia on the launch pad 39-A prior to the launch of STS-1 in April 1981. The distinctive black 'chines' visible on the upper surface of the forward wings are unique to this Orbiter.*

◀ *September 1976, Gene Rodenberry and cast members from the* Star Trek *television show line up beside NASA's Administrator Dr James D. Fletcher (left) to mark the roll-out of a new starship* Enterprise.

▶ Enterprise *made five free flights as part of the Approach and Landing Test (ALT) programme. The tail cone smoothed the airflow over the main engine area at the rear, and this was removed for the final two test flights.*

▶ *Rear view of* Discovery *with payload doors open during STS-120. The big nozzles of the three SSMEs are flanked by the two smaller ones of the Orbital Manoeuvring System (OMS).*

◀ Atlantis *after being de-mated from the Shuttle Carrier Aircraft (SCA) at the conclusion of the STS-125 mission.*

COLUMBIA OV-102

The first space-worthy Orbiter, *Columbia*, launched from Kennedy Space Center (KSC) on 12 April 1981. For the initial shakedown flights OV-102 was fitted with an ejector seat system for its two-man crew. *Columbia* weighed more than subsequent Orbiters by as much as 8,000lb (3,600kg). Wing and fuselage spars were heavier and it had an internal airlock which was not fitted to the others. It was the only Orbiter with an all-tile thermal protection

▶ Challenger photographed in orbit in June 1983. The picture was taken aboard the Shuttle Pallet Satellite which was later retrieved by the remote arm.

▶▶ The 'glass cockpit' or Multifunctional Electronic Display Subsystem (MEDS), as fitted on Atlantis – shown here in the simulator at Johnson Space Center.

system, although insulation blankets were later fitted to the fuselage and upper wing surfaces instead of the tiles. The first operational mission, STS-5, launched on 11 November 1982. *Columbia* completed a total of twenty-eight flights before its catastrophic destruction during re-entry in February 2003. (See 'Returning to Earth'.)

CHALLENGER OV-099

The second Orbiter into space, *Challenger*, had been built as a structural test airframe to test responses to heat and stress. Converted to operational status, OV-099 first launched with the STS-6 mission on 4 April 1983; a flight that saw the first spacewalk of the Shuttle program. With fewer tiles than its predecessor, *Challenger* could carry 2,500lb (1,100kg) more payload. The first Orbiter to launch at night, it took the first American woman into space and also made the first landing at KSC. Tragically it was destroyed in January 1986 during the launch of STS-51-L, its tenth mission. (See *'Challenger* – A Major Malfunction'.)

DISCOVERY OV-103

With thirty-six missions and 4,764 orbits to its credit – as of autumn 2009 – *Discovery* is the workhorse of the Space Shuttle programme. The third into space it is now the oldest Orbiter still in service. Upon its completion, in 1983, OV-103 weighed some 6,870lb (3,120kg) less than *Columbia*. Since then it has undergone two major upgrades, plus further safety modifications as part of the Return to Flight programme, following the *Columbia* disaster. It was *Discovery* that took the Hubble Space Telescope into orbit in 1990 and it has visited the *Mir*

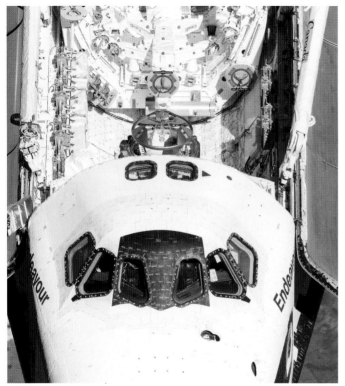

space station once, and the International Space Station (ISS) nine times. At the time of publishing it is scheduled to be the last Orbiter to fly, concluding with the STS-134 mission to the ISS in September 2010.

ATLANTIS OV-104

Completed in 1985, *Atlantis* is the lightest Orbiter so far. OV-104 made its first flight on 3 October 1985, carrying a classified defence payload on mission STS-51. To date it has completed thirty missions. *Atlantis* was the first Orbiter to dock with *Mir* and subsequently docked with the Russian space station a further six times, as well as making nine dockings with the ISS. It has also served as an in-orbit launch site for the planetary probes *Magellan* and *Galileo*, as well as launching the Compton Gamma Ray Observatory.

ENDEAVOUR OV-105

The final Orbiter, *Endeavour*, was commissioned following the loss of *Challenger* and first flew in May 1992 on STS-49. The use of leftover spare parts from the construction of *Discovery* and *Atlantis* made this a cheaper option than refitting *Enterprise*. OV-105 incorporated new hardware including improved plumbing and electrical systems, a 40ft (12m) drag chute for landings, updated avionics, improved nose wheel steering and Auxiliary Power Units (APUs). In theory these Extended Duration Orbiter (EDO) modifications facilitate longer missions of up to twenty-eight days. Further upgrades include a new electronic display system referred to as a 'glass-cockpit' and a new airlock, along with post-*Columbia* safety improvements.

PATHFINDER OVD-098

There is one other Orbiter which, although it never took to the air, earned itself an honorary OVD number. *Pathfinder* is a Space Shuttle simulator made of steel and wood, built to check roadway and structural clearances, although it is just slightly shorter than the real thing. In 1999, NASA borrowed back some of the forward SRB assemblies for the Shuttle fleet as the ones in use had been damaged or lost in operation. It is now on display in Huntsville, Alabama. (See Appendix 1 – 'Seeing the Space Shuttle'.)

A close-up view of Endeavour *as seen from the ISS. This final Space Shuttle Orbiter was named after the ship captained by the explorer James Cook, hence the English spelling.*

COUNTDOWN TO LAUNCH

A Shuttle launch entails years of detailed planning and several months of ground processing activities. Each mission has a specific launch window, a period normally lasting several hours, during which launch must occur in order to fulfil that mission's requirements. The factors determining this include the laws of orbital mechanics, the position of other spacecraft when a rendezvous is planned, the time of the landing and even the position of the sun in relation to critical mission activities.

After returning from space each Orbiter spends an average of four to six weeks in the Orbiter Processing Facility (see The Reusable Spacecraft). Once this process is completed it is taken to the Vehicle Assembly Building (VAB), which was built in the 1960s for the assembly of Apollo's Saturn V rockets. This is where it will be mated with the ET and SRBs. The VAB is big enough to process two Shuttles at any one time.

The refurbished SRBs and brand new ET typically arrive about six months before the scheduled launch day. Each SRB is delivered in four segments that must be

> *STS-98 leaves the Vehicle Assembly Building (VAB) on the start of its slow journey to the launch pad.*

'Every one of us is aware there is a slightly increased risk if you compare it to the day-to-day risk that we might be exposed to driving on the streets or going on commercial airlines. Each of us, independent of our nationality or space agency, believes the experience we gain in terms of scientific results, in terms of just expanding our horizons, is worth the remaining risk.'

German astronaut Thomas Reiter – prior to launch of STS-121, 25 June 2006.

> *Sections of a Solid Rocket Booster (SRB) during the stacking procedure inside the VAB.*

stacked one upon the other and fastened with connecting bolts to form the complete booster. A tight seal between the boundary or joint between each section is ensured with O-rings which are designed to prevent any seepage of hot gasses as the propellant is consumed.

With the twin SRBs standing vertically, the empty ET is lowered into place and connected to them. The Orbiter is then mated to the tank and bolted into position. Another relic of the Apollo era comes into play at this point. One of a pair of Mobile Launch Platforms (MLPs), or crawlers, transports the entire Shuttle on a 1mph (1.6km/h) journey along the 3.5-mile (5.5km) gravel crawlerway to the launch pad. This is known as the roll-out stage. There are two launch pads at KSC, 39A and 39B, and on rare occasions there have

The cylindrical Joint Airlock Module for the International Space Station (ISS), together with the Space Lab Double Pallet, are being prepared to be loaded within the Shuttle's payload bay for mission STS-104 in 2001.

been Shuttles waiting for launch on both of them. Once at the pad the Shuttle is stood on support posts and the MLP is removed. The Shuttle stands beside the steel launch tower, or Fixed Service Structure (FSS), which contains twelve access decks and a number of access arms extend to various parts of the spacecraft.

Most countdowns commence seventy-two hours before the launch and the time left is denoted as 'T minus' with T-0 being the actual moment of lift-off. However, this countdown allows for several hold periods, scheduled and unscheduled, during which time the clock will be stopped. At T-6 hours a sixty-minute pre-planned hold marks the time when the Launch Director will seek a favourable weather report before giving the go-ahead to the pad engineers to proceed with fuelling

the ET. This is usually completed by T-3 hours, although the tank will be constantly replenished to allow for boil-off of the cryogenic fuels. Any gas resulting from boil-off is released by external vents and can be seen in the pre-launch photos as plumes of white.

A two-hour hold occurs at T-3 hours and the astronauts are woken to be given a physical examination, eat their breakfast and clamber into the orange pressure suits. (On early Shuttle missions they wore blue overalls into orbit, but this was later changed to the high-visibility orange pressure suits.) They leave the checkout building shortly after the countdown resumes at T-2 hours 55 minutes and enter the famous crew transfer bus for the twenty-minute drive to the launch pad. Once there an elevator takes them up the tower to the 'White-

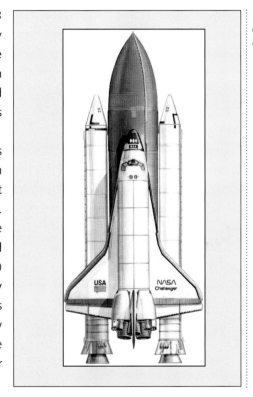

◄◄ Discovery *STS-124 during final mating to the ET's cradle.*

Did you know?
The Space Shuttle was the first manned spacecraft to use Solid Rocket Boosters (SRBs) which cannot be shut off and will continue to burn until all of the propellent is exhausted.

▶ The Discovery orbiter is being lowered into position for mating to the ET in preparation for mission STS-124.

▶▶ A relic of the Apollo era, one of the two MLP crawlers which originally cost US$14 million each. Weighing in at 2,400 tons, these formidable machines travel the 3.5 miles (5.6km) of the crawlerway at 1mph (1.6km/h) fully loaded.

room', situated at the end of the crew access arm, where technicians assist them in boarding the Orbiter and secure them in their seats. The hatch is sealed and the technicians depart, leaving the astronauts sitting on top of around 2,000,000lb of highly explosive rocket propellent.

Over the next two hours the Launch Control Centre carefully monitors the weather conditions, not just at the Cape but also at the abort landing sites. On the flight deck the commander and

'Anyone who sits on top of the largest hydrogen-oxygen fuelled system in the world; knowing they're going to light the bottom – and doesn't get a little worried – does not fully understand the situation.'

John Young – talking about the first Space Shuttle flight.

> On its way to the launch pad, roll-out of STS-63 on the Mobile Launch Platform (MLP), January 1995.

pilot verify operational readiness of the communications system, and prepare the primary flight-control systems.

At T-20 minutes the Launch Director conducts a final briefing for his launch control team and the countdown enters the last planned hold at T-9 minutes as he checks with each controller on the status of the major systems. All must give a 'go' for the launch to proceed. The count resumes. At T-7 minutes 30 seconds the crew access arm is retracted. T-6 minutes and the APUs that provide hydraulic power are warmed up. At T-4 minutes 30 seconds the Shuttle switches to internal power and the computer carries out a test gimbal of the engine nozzles before returning them to launch position at T-3.

T-1 minute 57 seconds, the oxygen vent access arm is retracted from the top of the ET.

◀ *Commander Scott Altman leads his fellow crew members of STS-125 from the transfer bus on the final walk to the lift at the bottom of the launch tower.*

▶ *White-room technicians make final checks and adjustments to the spacesuit of Dave Williams, a Canadian mission specialist, prior to entering* Endeavour *for mission STS-118.*

▶▶ Endeavour *STS-118 on the launch pad 39A at the Kennedy Space Centre (KSC), awaiting its latest trip to the ISS in August 2007.*

Did you know?
Built in the 1960s to accommodate Apollo's Saturn V rockets, the Vehicle Assembly Building (VAB) is the fourth largest building in the world by volume.

At T-31 seconds control of the countdown is passed over from the Ground Launch Sequencer (GLS) to the Orbiter's computers – a process known as auto sequence start. T-11 seconds and water floods into a trench below the launch platform as part of the sound suppression system which protects the Shuttle from acoustic shockwaves. T-8 seconds and sparks appear beneath the main engines. T-6.6 seconds, the main engines start sequentially at 120 millisecond intervals, causing the Shuttle to sway by as much as 6.5ft (2m). They must reach 100 per cent thrust within three seconds or the computers will initiate an automatic abort. Clouds of steam billow up around the launch pad as the hot exhaust from the engines hits the water trench. The computer verifies that the engines are functioning properly and the SRBs are ignited. Once started they cannot be turned off and as soon as they reach a stable thrust ratio, explosive bolts release the Shuttle. We have lift-off!

BURAN – THE SOVIET SHUTTLE

Not to be outdone, the Soviets wanted their own spaceplane. The creation of *Buran*, meaning 'blizzard' or 'snowstorm', can be viewed as a last mad act of Cold War paranoia resulting from Soviet insecurities – over the possible military use of the US Shuttle to bomb Moscow – and as a boost to national pride. Development work began in the 1970s and the first sub-orbital test flight of a scale-model vehicle took place in 1983. *Buran's* first, and so far only, orbital flight was on 15 November 1988 when the full-size *Buran*, OK-1K1, lifted off on the powerful *Energia* launch vehicle. Unmanned, it completed two orbits before making an automated descent for a runway landing at the Baikinour Cosmodrome.

Outwardly *Buran* appears to be a carbon-copy of the Space Shuttle, raising speculation that espionage may have come into play, but under the skin there are significant differences. *Buran* does not have its own engines and it rides piggyback on *Energia* as a passive payload. The 202ft (61m) tall *Energia* features four liquid hydrogen/oxygen rocket engines of its own, plus four reusable strap-on boosters fuelled with kerosene and liquid oxygen. *Buran* does have two jet engines for increased manoeuvrability within the atmosphere.

The project was terminated in 1993, by which time five production and eight test vehicles had been started or completed. (See Appendix 1 – 'Seeing the Space Shuttle'.)

LIFT-OFF!

Moving in what appears to be slow motion at first, the Shuttle takes an agonising three seconds to clear the launch tower. At this point the Launch Control Centre at KSC hands over to Mission Control at the Johnson Space Center in Houston, Texas. All three Main Engines (SSMEs) and the two slender SRBs are firing in unison to lift the 4,400,000lb (2,000,000kg) load off the ground. Riding

> *Flight controllers monitor their consoles at Mission Control at the Johnson Space Center in Houston, Texas.*

an incandescent firework, the Space Shuttle climbs upwards on streams of blazing fire, trailing a column of smoke from the SRBs and sending out a thunderous crackling wave of sound that splits the air and reverberates within the chests of spectators many miles away.

Shortly after clearing the tower the Shuttle begins a roll and pitch motion around its vertical axis, placing the Orbiter beneath the ET, and it ascends in a progressively flattening arc to set its orbital inclination and to take advantage of the Earth's eastward motion. Attaining low orbit requires more horizontal acceleration than vertical, and this manoeuvre is known by NASA as the 'roll and pitch-over'. Former Apollo 11 astronaut Mike Collins wrote that the Shuttle's ascent is much more rapid than the Saturn V, 'but still slow enough to register in eye and brain. The Shuttle's exhaust was almost too bright to watch. It was almost too beautiful to watch.' The vehicle continues to accelerate rapidly as the weight of the ET and the

The three Space Shuttle Main Engines (SSMEs) firing as Atlantis STS-110 rises from launch pad 39B on 8 April 2002.

▶ Discovery *STS-124*, lifting off on 31 May 2008.

▶▶ *The crew of STS-81 during a simulated countdown run-through at launch pad 39B.*

Did you know?
The first woman astronaut to pilot the Shuttle was Eileen Collins, on *Columbia* mission STS-93 in July 1999.

SRBs decreases as propellants are turned into raw power. Through the Orbiter's windows the crew see the sky turning from blue to navy to deepest black and by T+50 seconds the Shuttle has passed Mach 1, the speed of sound, and is already at an altitude of 5.7 miles (9.1km). At 'Max Q', the point when the aerodynamic forces reach their maximum, the main engines are

'It takes approximately eight and a half minutes of powered flight from lift-off on the pad until you're going 17,500mph. When you're on board that's a very exciting ride – goes very quickly. When you are sitting in a management position that is the longest eight and a half minutes in the world.'

Robert Crippen, veteran of four Shuttle missions and director KSC 1992–1995.

▶ *Aerial view of STS-2* Columbia *taken by fellow astronaut John Young aboard NASA's Shuttle Training Aircraft (STA), November 1981.*

Did you know?

The Shuttle's swaying motion just before launch, caused by the offset thrust of the three main SSMEs, is known as the 'nod' or 'twang'. It lasts for approximately six seconds and the top of the stack sways by as much as 6.5ft (2m) before settling back to the vertical.

momentarily throttled back to 67 per cent thrust to avoid over-stressing the Orbiter's structure.

At T+126 seconds the crew experience a sudden deceleration as explosive bolts are fired to release the SRBs and small separation rockets push them to either side away from the Shuttle to then fall back to the Atlantic Ocean. At the moment of separation the long smoke plume comes to an end, leaving only a fiery dot, no brighter than a distant sun, visible to those on the ground. The Shuttle is now at about 29 miles (46.3km) altitude. It is still climbing and begins accelerating to orbit using the thrust of the three SSMEs alone – a far smoother and quieter ride for the astronauts than the rough roller-coaster created by the SRBs. The loss of their thrust, however, means that the SSMEs have insufficient power to exceed the force of gravity and although the Shuttle continues to climb the rate of acceleration

◀ Explosive bolts release the Solid Rocket Boosters (SRBs) 126 seconds after launch. Small separation rockets push them away from the orbiter and they are parachuted back to the sea for collection.

▲ Discovery *STS-51 photographed in orbit by the SPA-ORFEUS satellite which it deployed and retrieved in September 1993.*

'A launch is very fast, and is violent and very noisy. It's basically vibrations and noise. I'm scared to death of launches … but I accept that risk.'

Story Musgrave, Space Shuttle astronaut.

decreases momentarily. Then, as the weight of the remaining liquid hydrogen/oxygen propellent reduces, the ever-lighter vehicle accelerates once more and at an incredible rate. It soon reaches speeds of around 12,400mph (20,000km/h), although this is still not enough to attain orbital velocity. At approximately T+7 minutes the Shuttle begins a shallow dive in preparation for ET separation and to gain velocity. In the final ten seconds of main engine burn the mass of the spacecraft has reduced to such an extent that the engines are throttled back in order to keep the rate of acceleration down to a more tolerable level for the astronauts.

At T+8 minutes 30 seconds, the SSMEs are shut down before the propellent is completely exhausted as running them dry would damage the engines. T+8 minutes 50 seconds, the ET is jettisoned by firing explosive bolts. Momentarily this rust-red torpedo seems to hang motionless against the marbled blue of the Earth before it tumbles backwards, to break up and burn high above the Pacific or the Indian oceans.

The twin Orbital Manoeuvring System (OMS) engines are fired to increase the perigee of the Shuttle's orbit; its nearest position to the Earth. This places the Orbiter in an optimum position should it have to return in the event of an OMS malfunction for example. It finally reaches its highest point, apogee, approximately 40–50 minutes after lift-off, and a further burn of the OMS places the spacecraft in a circular orbit at an altitude of around 186 miles (300km). This burn marks the last event of the ascent to orbit. The fireworks are over and the Space Shuttle has reached its destination – the largest and heaviest single vehicle ever placed into orbit.

These figures are within the parameters of a standard Space Shuttle launch and will vary from mission to mission.

CHALLENGER – A MAJOR MALFUNCTION

The mission patch for STS-51-L shows the two mission specialists at the bottom, and an apple symbol to denote the Teacher in Space program.

The Challenger *crew. Back row, left to right, Ellison Onizuka, Christa McAuliffe, Gre Jarvis and Judy Resnik. Front row, pilot Mike Smith, Commander Dick Scobee, and Ron McNair.*

At 11.38 EST on the morning of 26 January 1986, STS-51-L finally lifted off from launch pad 39B following a series of delays. After five years in service, Shuttle launches had become almost routine and it was only the presence among *Challenger's* crew of a young teacher from Concord, New Hampshire, that attracted widespread media and public interest in this one. Christa McAuliffe had been selected from 11,000 candidates to give the first school lessons from orbit as part of NASA's much heralded Teacher in Space Program. The mission patch even featured an apple for the teacher.

There had been the usual smiles and waves from the crew when they boarded the transfer bus several hours earlier. Wearing the blue overalls in which they would travel into orbit, they were unaware of the protracted debate during the night between NASA and the contractors who built the Shuttle. It had been an exceptionally cold night at the Cape with temperatures falling to -1°C, well below NASA's own guidelines for a safe launch. In particular, the engineers at Morton Thiokol, designers of the SRBs, were worried about the integrity of the rubber O-rings which sealed the joints between the booster's main sections. In the end they were overruled by their own management and their worries were not communicated adequately to the managers at KSC who were keen to restore the Shuttle's launch manifest. Unfortunately the run of forty-nine successful launches had engendered a culture of complacency that would cost them dearly.

During the night the Ice Team had also been hard at work, removing the

fingers of ice that smothered the launch pad. Engineers at Rockwell International, the Shuttle's prime contractors, were worried that if any ice dislodged during the launch it could damage the fragile thermal protection tiles. Accordingly the launch had been delayed for another hour while the ice began to melt.

As STS-51-L cleared the tower any concerns appeared to be unfounded, but almost immediately remote cameras recorded a problem. Puffs of dark smoke emanated from the right-hand SRB, near the strut that attaches the booster to the ET. These were caused by the opening and closing of a breech in the aft O-rings. The inner ring had hardened in the cold temperatures and the secondary outer ring was not seating properly as the tall booster flexed. At T+58.788 seconds a tracking

camera captured a plume of flame as hot gases escaped through the joint, burning into the ET. At T+68 seconds Mission Control gave the routine instruction 'Go with throttle up'.

'Roger, go at throttle up,' was the last communication from *Challenger's* commander, Dick Scobee. At around T+72 seconds the SRB pulled away from the aft strut, the ET was breeched and *Challenger* was immediately torn to pieces by the aerodynamic forces. From the ground it appeared to be an explosion with the SRBs curling away like twin devilish horns, although some spectators assumed this was a normal SRB separation. It soon became apparent that something had gone badly wrong when a NASA public affairs officer announced that flight controllers were looking very carefully at the situation.

◄ *A camera at launch pad 39B records the moment as STS-51-L climbs away. A puff of dark grey smoke can be seen coming from the aft joint of the SRB – lower right.*

◄◄ *Icicles draping the launch pad at KSC on the day of* Challenger's *launch.*

'Obviously a major malfunction.' The crew cabin survived the vehicle break-up intact, and it is thought that some of the seven astronauts may have been alive during the 65,000ft (19.8km) fall as some of the emergency air packs had been activated, but they were probably unconscious by the time the cabin hit the water.

In the aftermath of the *Challenger* disaster further Shuttle flights were

◀ *A USCG recovery team retrieves a piece of the* Challenger *Orbiter from the Atlantic a couple of days after the accident.*

◀◀ *The moment of* Challenger's *destruction with the trails of the SRBs entwined around a ball of gas from the external tank.*

'That six seconds seem to last for an eternity. The vibration was so high you can feel the Shuttle straining against the launch pad. You think it is going to shake itself apart. Then bang at zero when the solids light.'

Colonel Kevin Cholton, Shuttle pilot/commander.

suspended and a number of improvements were instigated. These included a redesign of the SRBs, a reduction in the unrealistic launch schedule and a revision of launch decision procedures and safety measures. In future all Shuttle crews would wear pressurised Advance Crew Escape Suits during ascent or descent, and additional options for evacuating the Orbiter or aborting during ascent, were devised.

'All of a sudden, space isn't friendly. All of a sudden, it's a place where people can die ... Many more people are going to die. But we can't explore space if the requirement is that there be no casualties; we can't do anything if the requirement is that there be no casualties.'

Isaac Asimov – commenting on the *Challenger* investigation, April 1988.

◁ *Kenneth Bowersox and Scott Horowitz of STS-82 practise an emergency escape from the launch pad in a slidewire basket.*

◁◁ *A crew member of STS-107 practises an emergency escape, or 'egress' in NASA parlance, using the escape pole.*

Did you know?
The Space Shuttle's main engines and rocket boosters can only propel it into orbit. When it returns to Earth it must land as an unpowered glider.

▲ Following a simulated emergency escape from the launch pad, the crew of STS-125 practise driving the M-113 armoured personnel carrier which will take them away to safety.

However, given the sudden catastrophic destruction of *Challenger*, none of these measures would have saved the crew of STS-51-L.

During launch and ascent the most likely abort scenario would be a main engine failure and there are several abort modes to deal with this. A Redundant Set Launch Sequencer Abort occurs between the start of the SSMEs at T-6.6 seconds but before T-0. This is under automatic computer control and has happened five times so far. Once the SRBs have been ignited it is a different matter entirely as they cannot be turned off, and it is only when they have burnt-out 123 seconds later that various 'intact' abort modes allow for a safe return of an Orbiter.

In a Return to Launch Site Abort (RTLS), the Shuttle continues downrange until the SRBs are jettisoned, and it then pitches around and heads back to KSC. The ET is jettisoned and the Orbiter makes a gliding landing. With a Transoceanic Abort Landing (TAL), the Shuttle continues to one of a number of designated landing sites in Africa or western Europe. (The long runway at RAF

Fairford, in Gloucestershire, is one of these.) An Abort Once Around (AOA) comes into play if the Shuttle has not achieved a stable orbit but still has sufficient velocity to circle the Earth once before landing. An Abort To Orbit (ATO) can be implemented if the Orbiter cannot achieve the intended orbit but can attain a lower one. This occurred with STS-51-F in July 1985.

Further contingency aborts exist where the Shuttle is unable to reach a runway and post-*Challenger* enhancements include a bailout procedure for such an event. There are no ejector seats on the Orbiter, and instead a hatch is blown and the crew must slide out on a pole, to clear the Orbiter's left wing, before descending by parachute.

All of these abort and escape procedures are designed to optimise the astronauts survivability in certain situations, but spaceflight is still a risky business.

Did you know?
In 1995 the launch of STS-70 was delayed after woodpeckers had bored holes into the insulation foam covering the external tank. Since then bird scarers and plastic owl decoys have been deployed around the launch tower.

LIVING IN ORBIT

Zero-gravity, or zero-g, is something of a misnomer as an orbiting spacecraft is still held by the gravity of the planet it is orbiting. Gravity has not been eliminated, but instead the spacecraft is in a state of freefall where the forces of acceleration and gravity are equalised. Weightlessness can be exhilarating, but it does come with its own set of unique problems for those who wish to live or work in space.

▶ A room with a view. Dan Tani on board Atlantis STS-122, with the International Space Station (ISS) visible through the window.

▶▶ Shane Kimbrough and Sandra Magnus, both STS-126 mission specialists, juggle fruit in Zero-G on the middeck of Endeavour.

Kathy Sullivan and Sally Ride examine an early version of a sleep restraint on Challenger *STS-41-G, in October 1984.*

'Space is fun. I quickly adapted to zero gravity and have had no problems eating or sleeping. It's fun to eat upside down and sleep on the ceiling.'

Bill Oefelien, STS-11.

In the short term it can make you feel sick and long-term weightlessness can cause significant adverse effects on the human body, including muscle atrophy, a deterioration of the skeleton, a slowing of the cardiovascular system, decreased production of red blood cells and a range of lesser symptoms.

The crew compartment of the Space Shuttle is fully pressurised and life support systems produce an atmosphere which resembles that on Earth at sea-level. Once the redundant crew seats have been stored away – you do not need to sit down in zero-g – it becomes a reasonably spacious working/living space. The mid-deck contains storage areas, as well as a personal hygiene station, plus the access to the payload bay. No special clothes are needed, shorts and t-shirts usually suffice, and unlike their space station counterparts the Shuttle crews have a fresh change of clothes for every day of their mission.

Eating can be a challenge in zero-g as you do not want liquids or loose food floating around. Some foods, such as fruit or brownies, can be eaten in their natural form while others, such as pasta, require the addition of water. The Orbiter has equipment to heat water and foods to

Did you know?
When US Senator Jake Garn flew on STS-51-D in 1985, he experienced such extreme space sickness that the 'Garn' became an unofficial measure for the highest possible level of the condition.

◄ *No one can hear you snore in space. Crew members of STS-120 get some sleep.*

Steve Frick, of STS-122, gets to grips with a floating snack on the mid-deck of Atlantis.

the proper temperature. Astronauts eat three meals a day and the nutritionists ensure that their food provides the right balance of vitamins, minerals and calories. Unlike the astronauts in the early days of the space programme, today's Shuttle and International Space Station (ISS) crews are able to choose from a wide range of foods and drinks with only a couple of provisos; bread is replaced by tortillas which take up less room, and salt and pepper are only available in liquid form because a powder would simply float away.

Just as on the Earth, astronauts need their sleep. With no up or down in weightlessness they can do this in any orientation, but some form of restraint is vital to stop them floating about the cabin. Most usually they use sleeping bags attached to the walls, and there are also four compact bunk beds available. Generally the schedule calls for eight hours sleep at the end of each mission day, although the excitement of being in orbit and the effects of space sickness can disrupt this sleep pattern. Space sickness, otherwise known as Space Adaptation Syndrome (SAS), is a common side effect of weightlessness and symptoms include nausea, vomiting, loss of appetite, dizziness and a general lethargy. It is estimated that almost 50 per cent of astronauts have suffered from this condition caused by the conflicting signals from the inner ear and the eyes reaching the brain. However, the human body usually adjusts to its new situation within a day or two at most.

During their time in orbit the Shuttle's astronauts adhere to Earth time. Each 'morning' Mission Control rouses the crew with a burst of wake-up music from a selection chosen by the astronauts or their family members, and the choice and style of music ranges enormously with individual tastes. Flight planners also schedule rest and relaxation breaks every day and one of the most popular activities is to stare out of the windows and watch the world go by. The Shuttle orbits the Earth every ninety minutes, bringing a new sunset or sunrise every forty-five minutes.

The average length of a space shuttle mission is too short to cause concern for the prolonged physiological effects of weightlessness, although for astronauts undertaking long-term missions on the ISS these may be more significant. They

▲ Steven Hawley runs on a treadmill on the mid-deck of Columbia STS-93 to help evaluate the treadmill vibration isolation system planned for the ISS.

'Old folks have dreams too, as well as young folks, and then work toward them. And to have a dream like this come true for me is just a terrific experience.'

John Glenn on his return to space on STS-95.

are minimised to some extent through a strict exercise regime which is achieved in weightlessness conditions via the use of treadmills, rowing machines and exercise bicycles.

NASA scientists are still studying the long-term effects of weightlessness and in 1998 they recruited a willing guinea pig in the shape of America's first man to orbit the Earth, John Glenn. In 1962 Glenn had shot to fame when he encircled the globe three times in the tiny one-man Mercury capsule,

◀ At the age of seventy-seven John Glenn became the oldest person to fly in space. He is shown here taking photographs from Discovery's aft flight deck during STS-95.

◀◀ John Glenn, astronaut and US Senator, was the first American to orbit the world aboard Friendship 7 in 1962. Forty-six years later he returned to space on STS-95.

Did you know?

America's first female astronaut in space was Sally Ride who flew on *Challenger* STS-7 in 1983. Two Soviet women had already beaten her to it by then.

▶ Discovery *STS-95 landing at KSC on 7 November 1998.*

Friendship 7, and at the age of seventy-seven he went back into space aboard *Discovery* as a mission specialist on STS-95. The primary purpose of this exercise was more than the hugely successful publicity stunt it proved to be, for it also provided a unique opportunity to study the effects of spaceflight on the elderly. It is hoped that a greater understanding of the detrimental effects of weightlessness, and of the ageing process itself, could provide clues in the treatment of Earthbound osteoporosis. Speaking from the Space Shuttle the jubilant high-flying pensioner repeated the words he had spoken from *Friendship 7*; 'Zero-g and I feel fine.'

WORKING IN SPACE

For almost thirty years the Space Shuttle has been NASA's primary means of maintaining a manned presence in space, undertaking a wide range of tasks including deploying, rescuing and even repairing satellites, launching interplanetary probes and the Hubble Space Telescope, conducting scientific experiments and delivering components, constructing, supplying and servicing the ISS. One of the initial roles envisioned in the commercial and military marketing of the Shuttle was its ability to take satellites into orbit, although the costs in comparison with expendable launch systems turned out to be not as favourable as originally anticipated. Even so, the Shuttle has proved to be far more than a simple delivery system and the continued presence of men and women in the harsh environment of space has been vindicated by a number of spectacular in-orbit 'fix-it' missions.

The simplest way for the Shuttle to deploy a satellite is to take it to the desired orbit using the Orbital Manoeuvring System (OMS), and release it from the open payload bay. However, the Shuttle's

Astronaut Ron Garan participates in the STS-124 second spacewalk which lasted over seven hours. He is shown installing television cameras on the Japanese Pressurised Module on the International Space Station (ISS) to assist with robotic arm operations.

STS-114, Soichi Noguchi, of the Japan Aerospace Exploration Agency, with Stephen Robinson in Discovery's airlock at the end of the mission's third session of EVA activity, August 2005.

'We proved that repairing satellites is a do-able thing. Satellite servicing is something that's here to stay.'

Robert Crippen, commander STS-41-C, after repairing the Solar Max satellite.

Orbiter only flies in a comparatively low orbit, approximately 200–250 miles above the Earth, while most satellites need to be positioned far higher up, typically at around 22,360 miles (35,790km). In order reach this they require an additional boost from a small rocket, known as an upper stage, which is attached to the satellite itself and fired remotely once the Shuttle is at a safe distance. During the 1980s the Payload Assist Module (PAM) was the main method, deploying satellites of up to 2,750lb (1,250kg) into a higher geosynchronous orbit launched vertically from within a shrouded cradle located in the payload bay. For heavier satellites a more powerful Inertial Upper Stage (IUS) two-stage rocket is used and this is transported horizontally in the payload bay because of its greater length.

To assist with the task of moving objects in or around the payload bay, the Shuttle is equipped with the Canadian-built 50ft (15.2m) robotic arm known as the Remote Manipulator System (RMS), or 'Canadarm', operated by an astronaut working inside the crew cabin at the rear of the flight deck.

A small cylindrical airlock located in the mid-deck section allows access to the payload bay so that space-suited astronauts can conduct spacewalks, or EVAs (Extra Vehicular Activity). In practice most Shuttle EVAs take place within the payload bay and the spacesuits provide adequate power and oxygen for about eight hours work. Weightlessness makes the experience more of a space-float than a spacewalk, and handrails and tether attachment wires run the length of the payload bay to prevent the astronauts from floating away.

◀ On Endeavour's first mission, STS-49 in May 1992, the primary objective was the capture, repair and re-deployment of the 4.5 ton Intelsat VI Communications satellite which had become stranded in an unstable orbit. This was the first three-person EVA.

(Components of the ISS have similar rails and attachment points.) Foot restraints at the end of the robot arm can also provide a secure mobile work platform, putting the

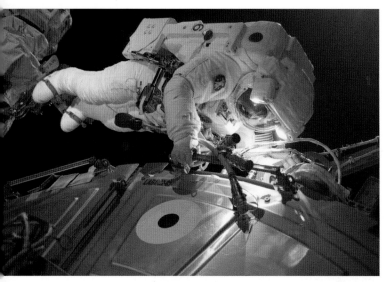

▲ *Darkness comes every ninety minutes and astronaut Piers Sellers is using artificial lighting to continue his work on the Destiny Laboratory during STS-112.*

astronaut into position when working on larger items.

Of course the Shuttle can recover satellites, sometimes making in-orbit repairs and correcting their orbits, or bring them back to Earth. Many satellites spin about their axis and an astronaut attempting to stop one by grabbing on will simply spin with it. A device known as a Manned Manouvering Unit (MMU) was developed to allow astronauts to safely fly free of the Orbiter without a cumbersome tether. By this means it is possible to attach to a spinning satellite's rocket nozzle and counter-act the spin using the MMU's nitrogen-powered thrusters. Once the satellite has been immobilised the robotic arm is then used to bring it into the payload bay. The first satellite to be captured like this was Solar Max during the STS-41-C mission in April 1984.

In May 1992 the crew of STS-49 intercepted the Intelsat IV telecommunications satellite which had become stranded in an unstable orbit. This 4.5-ton cylinder defied

◀◀ *James Newman floats about* Columbia's *payload bay during the STS-109 mission to repair the Hubble Space Telescope (HST), the base of which can be seen behind his head, March 2002.*

◀ *In preparation for an underwater simulation for EVA training, Richard Arnold of STS-119 is about to be submerged into the tank at the Neutral Buoyancy Laboratory (NBL) near the Johnson Space Center.*

initial attempts to grab it and it took three astronauts to finally move it safely into the open payload bay. They then attached a booster stage to its underside before it was released.

STS-51A was the last mission in which the MMU was used because post-*Challenger* NASA had become far more cautious about

▶ *These underwater training simulations provide astronauts with the nearest experience to working in weightless conditions during an EVA.*

▶▶ *Out there: On 11 February 1984, STS-41-B mission specialist Bruce McCandless became the first astronaut to move untethered away from his spaceship. Using the Manned Manoeuvring Unit (MMU), a nitrogen jet propelled backpack, he went free-flying to a distance of 320ft (97.5m) away from Challenger.*

'That may have been one small step for Neil, but it was a heck of a big leap for me.'

Bruce McCandless on making the first untethered spacewalk with the MMU.

◁ During STS-64, Mark Lee floats freely without a tether as he tests the new Simplified Aid for EVA Rescue known as SAFER.

Did you know?
Each permanent ISS crew is given a sequential expedition number, Expedition 1 and so on, and they have an average duration of around six months in orbit.

STS MISSION NUMBERING

NASA's system for numbering Shuttle missions requires an explanation: At first the numbering was sequential for the first nine missions from STS-1 onwards, simple enough. Then starting in 1984 each mission was assigned a code with the first digit indicating the fiscal year, the second designating the launch site ('1' was for KSC, '2' was for the USAF launch facility at Vandenburg AFB, even though this was never used), and a final letter at the end indicated the order of the launches. To further complicate matters these codes were assigned when the missions were originally scheduled, but in some cases the actual launches were delayed. This numbering system was used from STS-41-B through to the final *Challenger* flight STS-51-L. When flights resumed the missions were given sequential numbering from STS-26 onwards, but as before the actual order of the launches is often wildly out of sync. For example, STS-121 launched in July 2006, more than a year before STS-120 which flew in October 2007.

putting astronauts at risk. It has been replaced by a more compact device, the Simplified Aid For Extra-vehicular Acivity (SAFER), which incorporates some of the benefits of the MMU into a propulsive unit that attaches to the standard EVA backpacks. This self-rescue device was first tried out in space with an untethered spacewalk during STS-64 in September 1994.

Training astronauts to work in weightless conditions has always been problematic. NASA's zero-g aircraft can simulate

Rex Walheim, mission specialist on STS-122, works on the outside of the International Space Station's Columbus laboratory module.

Did you know?
Two Space Shuttle missions have been funded by the German government, Spacelab D-1 (for Deutschland-1) on STS-61-A, and Spacelab D-2 on STS-55. These had their own scientific mission controllers based at Oberpfaffenhofen, near Munich.

weightlessness in short bursts, by flying steep parabolic curves on a stomach-churning roller-coaster ride that deservedly earns it the nickname 'vomit-comet', but this method is not suited to protracted sessions. These are conducted in an overgrown fish tank, known as the Neutral Buoyancy Laboratory, which is located near to the Johnson Space Center in Texas. Here a fully kitted astronaut, complete with an underwater version of SAFER, can practise EVAs, moving about full-scale representations of the payload bay and the satellites or ISS modules and components they are to encounter in orbit.

No Earth-bound simulation can ever come anywhere close to the real thing. As the British-born astronaut Michael Foale once stated, 'I love EVA. I think it's a beautiful experience to be out there in a suit with just a faceplate between you and the vacuum. The Earth is an even richer colour than you can imagine. It's a fantastic view.'

◀ A fellow astronaut working in Discovery's *payload pay is reflected in Piers Sellers's mirrored visor during the third spacewalk of STS-121. Note the adjustable lights on either side of his helmet.*

SPACELAB AND SPACEHAB

Mission patch for Spacelab 1 which flew on STS-9 in November 1983.

Artist's impression from 1981 of the Spacelab laboratory module to be carried in the Orbiter's payload bay. Equipment such as telescopes, antennas and sensors, can be mounted on pallets for direct exposure to space.

Spacelab was not an independent spacecraft, but a collection of modular components which could be arranged in different combinations within the Space Shuttle's payload bay to form a series of tailor-made space laboratories. Instigated in 1973, it was a major multinational space project involving ten member nations within the European Space Agency (ESA) – formerly known as the European Space Research Organisation – that designed, constructed and financed it, together with NASA who provided the ground facilities and managed the Spacelab flights.

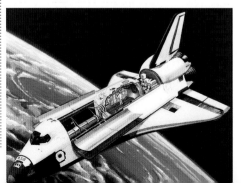

'Astronauts and teachers actually do the same thing. We explore, we discover, we share. The great thing about being a teacher is you get to do that with students, and the graet thing about being an astronaut is you get to do it in space.'

Barbara Morgan, STS-18 teacher-astronaut and backup to Christa McAuliffe, on fulfilling her legacy

The pressurised module was available in two segments. The core segment contained various support systems, such as data processing equipment, and was kitted out with floor-mounted racks and a workbench. The other segment was known as the experiment segment and provided additional laboratory space. Each module was cylindrical to fit the Shuttle's payload bay and measured 13.5ft (4.1m) in diameter

by 9ft (2.75m), plus a blunt end cone at either end. These modules provided a shirt-sleeve working environment for the mission and payload specialists who were nominated by the scientists and bodies sponsoring the experiments. They were trained by NASA. Up to four Spacelab specialists could be accommodated on a mission and they were usually divided into two teams in order to operate around the clock.

Because of centre-of-gravity issues with the Orbiter, Spacelab's habitable modules could not be located right up against the crew cabin, and instead a z-shaped aluminium tunnel connected the Shuttle's mid-deck airlock to the Spacelab located centrally within the payload bay. Un-pressurised U-shaped open modules, known as pallets, were usually arranged at the back of the payload bay and they carried research equipment and experiments that required an open view or needed to be exposed to the radiations and vacuum of space. Each pallet was 10ft (3m) long. The pressurised modules were constructed by an industrial consortium headed by ERNO-VFW Fokker, while the U-shaped pallet sections were built by British Aerospace.

◄◄◄ *Endeavour, docked to the Destiny laboratory of the ISS during STS-118 in Aug 2007. The Spacehab pressurised module is visible in the payload bay.*

◄◄ *Astronaut Mae Jemison working in the Spacelab-J module on board STS-47. The objectives of this 1992 mission included life sciences, microgravity and technology research.*

▲ *Kalpana Chawla shown reviewing science data in the Spacehab double module aboard* Columbia, *STS-107.*

▶ *Looking down at the payload bay containing the single Spacehab module on STS-118. A tunnel connects the pressurised module with the crew cabin of the orbiter.*

▼ *Diagram showing the arrangement of the Spacehab research double module on STS-107. The extended duration kit carries additional oxygen and hydrogen for the electricity-producing fuel cells.*

STS-107/Research-1 configuration

Tunnel Adapter | Spacelab Transfer Tunnel | Spacehab Research Double Module | Gas Bridge Assembly | Orbital Acceleration Research Experiment | Extended Duration Orbiter kit

The Spacelab experiments and studies covered a range of scientific areas including materials and pharmaceutical processing in zero-g, astronomical investigations and plasma investigations. In various configurations the Spacelab components flew on twenty-five Shuttle missions from Spacelab-1 on STS-9 in 1983, up to STS-90 in 1998. After that a pallet was recommissioned in 2002 for a handful of missions, but in general the plan had always been for the scientific work to be transferred to the ISS or to the new Spacehab.

Did you know?

Cosmic golf: in 2006 cosmonaut Mikhail Tyurin hit the longest golf drive in history sending a lightweight ball off into space during an EVA from the ISS.

> 'Of course risk is part of spaceflight. We accept some of that to achieve greater goals in exploration and find out more about ourselves and the universe.'
>
> Lisa Nowak – STS-121 astronaut, a few days prior to launch, June 2006.

◀ STS-107, with Michael Anderson holding a procedures checklist while working in the Spacehab Research Double Module. On 1 February 2003 the crew of Columbia was lost during re-entry and this photograph was processed from recovered film.

◀ Endeavour STS-118, with the Spacehab logistics module in the payload bay, August 2007.

Spacehab was a system of pressurised carrier modules, similar to Spacelab in many respects. They were manufactured by the Spacehab company (now Astrotech Corporation) which provides commercial space products and services to customers including NASA and the US Department of Defense. The Spacehab modules and integrated cargo-carriers provided supplemental ferrying capabilities to several Shuttle missions, including eight re-supply missions to the ISS, and seven to the Russian space station, *Mir*. The inaugural flight of Spacehab's manned Research Double Module (RDM) took place on *Columbia*'s ill-fated STS-107 mission in 2003.

HUBBLE TROUBLE

Technicians examine Hubble's 8ft (2.4m) mirror. This will reflect the light from distant stars, bounce it off a second mirror and it passes through a hole where instruments wait to capture it. The hole is covered up in this photo.

The Hubble Space Telescope (HST) is arguably the greatest advance in optical astronomy since Galileo invented the telescope. Named after American astronomer Edwin Hubble, the HST is a collaborative project between NASA and the European Space Agency (ESA), started in the 1970s with a view to launching in 1983. Unfortunately the project was beset by technical issues and the delays caused by the loss of *Challenger* in 1986.

Hubble was the only one of NASA's four 'Great Observatories' designed to be serviced by the Space Shuttle. Over a period of sixteen years there were five Servicing Missions (SMs) to Hubble and consequently, its capabilities grew immensely. Compared with terrestrial telescopes, Hubble is not particularly large. The primary mirror is 94.5 inches (2.4m) across and overall Hubble is about the size of a single-decker bus. However, the combination of precision optics and state-of-the-art instrumentation high above atmospheric distortions, provide it with an unparalleled clarity of vision. At the thin end of Hubble's silvery tube is the aperture which is shielded by a flap or door

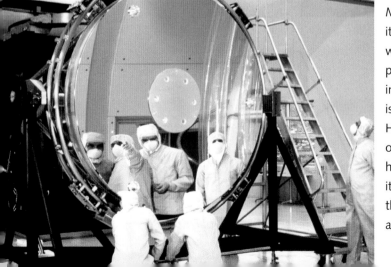

 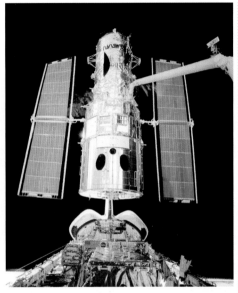

◄◄ *Hubble is lifted upright for testing in Lockheed Martin's acoustic vibration chamber in preparation for its 1990 launch aboard* Discovery *STS-41.*

◄ *Held by the robotic arm, the Hubble Space Telescope is unberthed from* Discovery's *payload bay during the second servicing mission, STS-82.*

to keep unwanted light out. Situated within the wider section, about three quarters of the way from the aperture, is the primary mirror which bounces light off a secondary mirror and back through a central hole to the instrumentation.

When launched Hubble had five main instruments: The Wide Field and

Planetary Camera (WF/PC), a high-resolution imaging device intended primarily for optical observations with its processing power divided between the Wide Field Camera and the Planetary Camera; the Goddard High-Resolution Spectrograph (GHRS), designed to operate in the ultraviolet range; the High-Speed Photometer (HSP), optimised for visual and ultraviolet observation of variable stars and

'The Hubble's successful launch will, I feel, be the most important piece of work NASA has done in recent years, and one that I hope will herald the agency's return to the forefront of science and exploration. It will be an institutional as well as a scientific milestone.'

Michael Collins, CSM pilot Apollo 11

other objects with variable brightness; the Faint Object Camera (FOC); and the Faint Object Spectograph (FOS).

On 24 April 1990, the Hubble Space Telescope finally rode into space aboard *Discovery* STS-31, and the following day it was released into low-orbit. Unfortunately it soon became clear that Hubble was seriously flawed. Hubble's primary mirror was the most precisely figured mirror ever made, but unfortunately it had been ground to a very precise but incorrect shape. The difference might be less than the thickness of a human hair, about 2.2 microns, but it was enough to cause a spreading of light and a blurring of the images. The all-seeing eye would need several adjustments to put this right.

In December of 1993, more than three years after Hubble's arrival in space, the first of the Service Missions, SM-1, was carried out by the astronauts of *Endeavour* STS-61. During five lengthy EVAs they replaced the High Speed Photometer with an optics package called the Corrective Optics Space Telescope Axial Replacements (COSTAR), consisting of two mirrors, one of which was configured to correct the telescope's troublesome eyesight. The WF/PC was also

◀◀ *European Space Agency (ESA) mission specialist Claude Nicolliere, works at a storage enclosure during the second of three STS-103 EVAs to service Hubble.*

◀ *Richard Linnehan uses a laser ranging device to measure the distance between* Columbia *and Hubble during the STS-109 mission in March 2002.*

⬩ Astronauts John Grunsfeld and Richard Linnehan work in tandem to replace one of Hubble's solar arrays during STS-109. Linnehan is standing on a foot restraint at the end of Columbia's robotic arm.

exchanged, by the WF/PC2 which features an internal optical correction system. The solar arrays and their drives, four gyroscopes used to point the telescope, electrical control units and other components were all replaced. The onboard computers were upgraded and Hubble was boosted into higher orbit to compensate for three years of drag in the thin upper atmosphere. The fix worked. An outstanding success for the supporters of the Shuttle programme and man's role in space.

Four further servicing missions followed. In February 1997, SM-2 saw *Discovery* return to Hubble on STS-82 to replace various items including the GHRS with the Space Telescope Imaging Spectograph (STIS), and the FOS with the Near Infrared Camera and Multi-Object Spectrometer (NICMOS). The thermal insulation was repaired and the telescope was given another nudge back into orbit. In December 1999, *Discovery* was back sooner than anticipated with STS-103 on SM-3A, fitting new gyroscopes after four out of six had failed, as well as replacing a fine guidance sensor, a computer, and the thermal insulation blankets. In March 2002, SM-3B saw *Columbia*'s STS-109 astronauts install an Advanced Camera System (ACS), restore NICMOS which had run out of coolant, and fit improved smaller solar arrays which provide more power – enabling all

'Spacewalking is just about the coolest and most rewarding thing an astronaut can do.'
Mike Massimino, STS-15 mission to Hubble.

◀ *Mission controllers monitor progress during the* Atlantis *STS-125 mission to Hubble in May 2009.*

Did you know?
The Space Shuttle's windows are made of aluminium-silicate glass and fused silica glass. They consist of an internal pressure pane, a 3-inch (7.5cm) optical pane and an external thermal pane.

◀ *The Hubble Space Telescope seen from the departing* Atlantis *on STS-125.*

instruments to run simultaneously – and reduce the Hubble's drag.

The final scheduled Shuttle mission, SM-4, took place with the launch of *Atlantis* STS-125 on 1 May 2009 with a number of replacement parts designed to extend the space telescope's useful life. Over the course of five EVAs astronauts installed the Wide Field Camera 3 (WFC3) and the Cosmic Origins Spectrograph (COS) – replacing COSTAR which had been fitted in 1993 to correct Hubble's eyesight. They also repaired the ACS and STIS and replaced other components including the nickel-hydrogen batteries which provide electrical power when Hubble falls within the Earth's shadow. As a result it is expected that Hubble will remain operational until at least 2014.

During its lifetime Hubble has made 800,000 observations, snapped over 500,000 images and produced 30 terabytes of data – the equivalent of 25 per cent of the information stored in the entire US Library of Congress. It has orbited the Earth more than 100,000 times clocking-up an incredible 2.4 billion miles. The intention had been to bring Hubble back to Earth on a Shuttle, but the retirement of the fleet makes this impossible and at the end of its service life it will be de-orbited to burn up in the atmosphere. There are sure to be other even larger space telescopes in the future, but none is likely to recapture the drama of Hubble's trials and tribulations or the impact of its incredible images of the far reaches of our expanding universe.

Hubble has provided unprecedented views of the far reaches of space. This is a bubbly ocean of glowing hydrogen and other elements in the Omega/Swan Nebula, a hotbed of star formation.

Did you know?
At one point the US Air Force had planned on having its own fleet of Shuttles which would have operated on defence mission from their own space launch complex at Vendenberg AFB, California.

SPACE STATIONS

The International Space Station (ISS) represents a colossal leap in cooperation among former Cold War and 'space race' rivals, in particular the USA and Russia. Both nations had been working on their own projects to establish a permanent facility in orbit with the Soviets/Russians taking an early lead with *Salyut*, a series of nine single module space stations operating between 1971 and 1982, and the bigger *Mir* which was continually occupied for a period of fifteen years from 1986 until 2001.

The USA had similar aspirations but a number of proposed projects fell by the wayside. Their first space station to reach orbit, *Skylab* which operated with three crews in 1973 and 1974, was cobbled together from leftover components of the curtailed Apollo programme. A decade later, US president Ronald Reagan announced

'We can reach for greatness again. We can follow our dreams to distant stars, living and working in space for peaceful, economic, and scientific gain. Tonight, I am directing NASA to develop a permanently manned space station, and do it within a decade.'

President Ronald Reagan, State of the Union address, January 1984.

▶ Mir *photographed from STS-79 in September 1996.*

▶▶ *Cosmonaut Valeri Polyakov looks out of* Mir*'s window during a rendezvous with* Discovery *STS-63 in February 1995 in preparation for the first Shuttle,* Mir *docking later that year.*

the Space Station *Freedom* program; in part to rival *Mir* but also as a direct result of the Space Shuttle entering service. With increasing budgetary constraints *Freedom* never made it into space, and with the fall of the Soviet Union it became obvious that only an international collaborative effort could undertake such a large and sustained endeavour. In 1992 the Americans and the Russians signed an agreement concerning cooperation in the exploration and use of space for peaceful purposes, in particular the construction of the ISS.

In preparation for this collaborative venture the USA became increasingly involved in the *Mir* programme with the exchange of a number of astronauts between the two nations. On 29 June 1995, *Atlantis* (STS-71) became the first Shuttle to dock with *Mir*, and they were locked together for five days. It delivered a relief crew of two cosmonauts to the space station, and afterwards returned one US astronaut and two Russian cosmonauts back to Earth. *Atlantis* made a further

Did you know?
Space debris ranging from spent rocket stages and defunct satellites to flakes of paint and lost tools, pose a constant threat to the ISS. The debris is monitored from the ground and the station's orbital altitude can be altered. On two occasions the crew have retreated to a *Soyuz* spacecraft in case an evacuation was required.

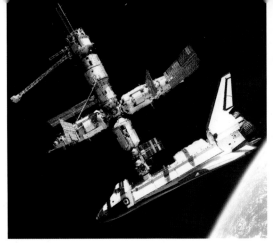

▲ Atlantis *STS-71* makes the first docking with Mir on 4 July 1995, photographed by the Mir-19 crew aboard a brief fly-around in their Soyuz *spacecraft*.

six dockings over the next two years, *Endeavour* made one in 1998 and the final Shuttle-*Mir* docking was *Discovery* STS-91's in June 1998. By the time *Mir* was deliberately de-orbited in 2001, assembly of the ISS was well under way. The first Shuttle mission for the ISS had taken place in December 1998, when *Endeavour* STS-88 joined the American *Unity* node to the Russian *Zarya* module which had been launched aboard an unmanned Proton-K two weeks earlier. *Unity* would provide berthing facilities for additional modules while *Zarya* supplied electrical power, storage and guidance during the initial assembly phase.

As of October 2009, *Endeavour*, *Atlantis* and *Discovery*, have made twenty-eight flights to the ISS between them. These have involved the delivery of several additional modules, including the *Destiny* research laboratory, the *Quest* primary joint airlock, *Harmony* which acts as the utility hub of the ISS and as a central connecting point, the *Columbus* laboratory built and operated by the European Space Agency (ESA), and the two sections of the Japanese *Kib* laboratory module. Further payloads

◀ *STS-71 mission commander Robert Gibson shakes the hand of Mir-18 commander Vladimir Dezhurov after their historic first docking.*

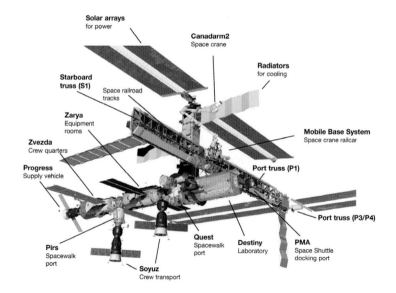

◀ *Diagram showing the major components of the International Space Station (ISS).*

▶ *The* Quest *Joint Airlock being moved into position after delivery by STS-104 in July 2001.*

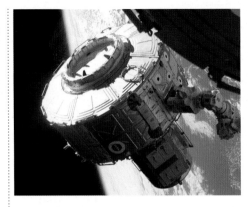

'The exterior of the Space Station looked like new. It was shiny and clean. It was easy to move around the exterior as there were lots of handholds along the way. Our long safety tethers were firmly attached to the structure, and if that failed we had our SAFER jet backpack to fly us back to the Station.'

Carl Walz, astronaut STS/ISS.

▲ *STS-88 was the first Shuttle ISS assembly flight, taking Node 1 into orbit in December 1988.*

are scheduled for the Shuttle including: STS-130 with the third and last of the USA's nodes, *Tranquility*, which will contain life support systems to recycle waste water for crew use and generate oxygen, as well as providing four additional berthing locations, and the *Cupola* observation module. STS-132 will carry the Russian-built Rassvet Mini-Research Module while STS-133 will see the final Shuttle assembly flight to the ISS with the European/American Permanent Logistics Module.

Space Shuttles have delivered many other components to the ISS including a robotic arm, Canadair-2, plus the external linking truss structure and solar panel arrays. Their crews have worked extensively on the assembly, maintenance and construction

Delivered by Atlantis *STS-122 in February 2008, the European Laboratory* Columbus *was the European Space Agency's largest contribution to the ISS. This cutaway of the 23ft-long (6.9m) module gives an idea of scale of the internal working space within the space station.*

Did you know?

In 1985 *Challenger*'s astronauts tested a special zero-g Coca-Cola can. An improved version was tried by Russian cosmonauts aboard *Mir* in 1991.

of the ISS and the Shuttles have provided the transportation on a number of crew rotations, taking astronauts of different nations to and from the space station. Of course, in this collaborative venture, the Shuttle has not been the only spacecraft to visit the ISS. Other vehicles have been utilised including the Russian *Soyuz* and unmanned *Progress* supply spacecraft, ESA's Automated Transfer Vehicle (ATV)

Discovery *STS-120* rendezvous with the ISS prior to docking in October 2007. The Harmony *mode is visible in* Discovery's *payload bay.*

and the Japanese H-II Transfer Vehicle. In the gap between the planned cessation of the Shuttle programme and the operational debut of its successor – see 'End of an Era' – only *Soyuz* will be available to deliver and collect ISS astronauts from orbit.

For a while it looked as though the Shuttle might be joined by another spaceplane. In the 1990s work had begun on the X-38, a collaborative project between NASA, ESA and the German space agency, to develop a space-lifeboat capable of bringing ISS crew members back to Earth in an emergency if an alternative was not available. Its design was a throwback to the lifting-body concept of the 1960s, taking the Shuttle concept full-circle. The X-38 would make a re-entry much like the Shuttle itself and then glide beneath a steerable parafoil for final descent and landing. The prototype first flew in 1999, but the project was cancelled in 2002.

The ISS is the largest man-made structure in space and it can be seen from the Earth with the naked eye. Its critics are quick to point out that the space station is also the most expensive object ever built with costs estimated at up to a staggering US$160 billion. During its time in orbit the ISS has been continually manned since November 2000. A host of experiments have been conducted across many fields including space medicine, life sciences, physical sciences and Earth observation. It has also provided a location for studies on the effects on the human body of long-term exposure to weightlessness, and has served as a test-bed for developing reliable spacecraft systems for the future. In years to come, space farers may look back to the ISS as a stepping stone to the stars.

◀ *Shuttle crew pose with members of the ISS crew on the last day of Endeavour STS-123's mission to the space station in March 2008.*

Did you know?
With astronauts coming from countries in different time zones, and orbiting the Earth sixteen times a day, the ISS adheres to Coordinated Universal Time (UTC) which is otherwise known as Greenwich Mean Time (GMT).

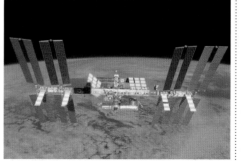

◀ *Like a huge dragonfly spreading its delicate wings, the ISS is the largest object ever assembled in orbit. Photographed by the crew of* Discovery *STS-119 in March 2009.*

RETURNING TO EARTH

Getting a spacecraft into orbit is only half the story. The process of getting one back down to Earth safely is fraught with technological challenges and danger. As it re-enters the atmosphere a spacecraft begins to push through the thicker air causing friction against the air molecules which generates intense heat. This process also creates a sheath of ionised air which blocks radio signals to and from the spacecraft for a period of around twelve minutes. Prior to the Shuttle, spacecraft such as the Apollo capsule survived the plunge through the atmospheres, thanks to an ablative heat shield. This functioned through a process known as 'blowing' by which a layer of the heatshield charred and melted during re-entry. This works fine for spacecraft that were for once-only use, but for the Shuttle a more sustainable system was required.

Thousands of small silica-glass Thermal Protection Tiles cover approximately 70 per cent of the Orbiter's exterior, and these 20,000 tiles, known as High-temperature Reusable Surface Insulation (HRSI), protect

> Discovery rests on the runway at Edward's AFB, California, after an early morning landing on 9 August 2005. STS-114 marked the return to flight after an eighteen-month delay following the loss of Columbia.

The Space Shuttle's dramatic descent through the atmosphere as depicted by a NASA artist.

atmosphere. Each tile is approximately the size of a chunky paperback book, with a thickness ranging from 0.4 inches (1cm) to 3.5 inches (9cm). Another material, the areas likely to encounter the greatest heat. Coloured black they are easily identified in the photographs covering the underside and the nose, and they can resist temperatures of up to 704°C by radiating most of the absorbed heat back into the

'You guys are doing a fantastic job staying on the timeline and accomplishing great science ... There is one item that I would like to make you aware of for the upcoming PAO event ... This item is not even worth mentioning other than wanting to make sure that you are not surprised by it in a question from a reporter ... Experts have reviewed the high speed photography and there is no concern for RCC or tile damage. We have seen this same phenomenon on several other flights and there is absolutely no concern for entry.'

Email sent to *Columbia* STS-107 regarding concerns about the foam strike. PAO refers to the NASA Public Affairs Office. 23 January 2003.

Did you know?
Unlike conventional aircraft, the Space Shuttle Orbiters do not have any external anti-collision, navigation or landing lights. All landings at night are made with floodlighting on the runways.

Columbia *making an early morning landing at KSC at the end of an earlier mission, STS-80, on 7 December 1996.*

called reinforced carbon-carbon, protects the leading edges of the nose and wings from temperatures of up to 1,260°C.

Re-entry is initiated when the Orbital Manouvering System (OMS) is fired in the opposite direction of motion to slow the Shuttle and bring it into a lower elliptical orbit within the upper atmosphere. This three-minute burn takes place approximately half way around the world from the landing site. The Orbiter is then flipped over and with its nose pointing upwards slightly and it encounters an increasing number of air molecules which slow its forward progress and causes the descent rate to increase. At 400,000ft (120km) it is travelling at twenty-five times the speed of sound, 18,000mph (30,000km/h). A nose-up attitude of 45° increases drag and minimises heating. As the air becomes denser the Shuttle begins to perform more like an aircraft than a spacecraft, and a series of four S-shaped banking turns dissipates some of the speed sideways to prevent the descent angle from flattening out too much.

As it levels out the nose is lowered into a shallow dive for the unpowered glide to the landing site. The descent rate is still very high at over 110mph (180km/h) with speeds of up to Mach 3. At an altitude of 9,800ft (3,000m) the pilots apply aerodynamic braking to slow it down and by touchdown the speed is around 215 to 425 mph (345–680km/h). With landing gear lowered the rear wheels screech as contact is made with the runway and the drag chute is deployed. Once the Orbiter has come to a stop it is left standing on the runway for several minutes to allow the poisonous hydrazine fumes from the thrusters and APUs to dissipate before the astronauts can disembark.

Whenever possible the Orbiter lands at the Shuttle Landing Facility at KSC, but several alternative landing sites exists if the weather condition are unsuitable at the Cape.

THE *COLUMBIA* DISASTER

On 1 February 2003, the Space Shuttle *Columbia* disintegrated over Texas during re-entry at the end of mission STS-107. It subsequently emerged during the investigation into the disaster that eighty-two seconds after launch a piece of insulation foam from the ET had broken

▲ *Mission patch for STS-107, with the central element depicting the 'μg' for microgravity flowing into the rays of the astronaut symbol.*

◀ *These terrible images of the break-up and destruction of* Columbia *during re-entry were flashed around the world on 1 February 2003.*

> *Members of a special return to flight task group examine some of the tiles recovered from the* Columbia *debris. Standing in the centre is former Gemini and Apollo astronaut Thomas Stafford.*

> *New methods for repairing the thermal tiles in orbit were tested in simulated Zero-g aboard a NASA aircraft.*

loose and struck the leading edge of the left wing, damaging the carbon-carbon thermal protection. The loss of insulating foam during launch was not-uncommon and despite the concerns of some engineers, NASA's managers failed to recognise the relevance of the damage to *Columbia* and even limited investigation on the basis that there was little they could do about it anyway. During re-entry the breach in the wing permitted hot gasses to penetrate the vulnerable internal wing structure, causing the rapid destruction of the vehicle and the loss of all seven astronauts.

As a result, the remaining Shuttle fleet was grounded for two years and the construction of the ISS was put on hold. The *Columbia* Accident Investigation Board addressed a number of technical and organisational issues. It concluded that a

A major focus of STS-114, in August 2005, was the evaluation of new techniques to inspect and repair the thermal tile heat shield. This photo was taken by one of Discovery's astronauts from the end of the robotic arm.

rescue mission might have been possible either by launching *Atlantis* (which was already being prepared for a 1 March launch) and keeping *Columbia* in orbit for a longer period, or through emergency repairs carried out by the astronauts. Unfortunately *Columbia* did not have a robotic arm which made inspecting the thermal protection system difficult, but not

The crew of Columbia STS-107, from left to right: David Brown, commander Rick Husband, Laurel Clark, Kalpana Chawla, Michael Anderson, pilot William McCool, and Israeli payload specialist Ilan Ramon.

'Some say that we should stop exploring space, that the cost in human lives is too great. But *Columbia*'s crew would not have wanted that. We are a curious species, always wanting to know what is over the next hill, around the next corner, on the next island. And we have been that way for thousands of years.'

Stuart Atkinson – 'New Mars,' March 2003.

impossible, and there still remained the question of how repairs would have been made using the materials available.

Despite its many successes, the Space Shuttle programme has seen two catastrophic accidents resulting in the loss of the *Challenger* and *Columbia* crews. If nothing else their sacrifice serves as a sober reminder of the very real risks involved in space exploration.

Did you know?

Most of the Space Shuttle's re-entry is performed under computer control, although it can be flown manually in the event of an emergency. While the approach and landing phase can also be controlled by the autopilot, it is usually flown by the crew.

➤ *Clad in thermal protection suits, rescue crew aid a volunteer crewman during a post-landing emergency egress drill at NASA's Dryden Research Center.*

THE REUSABLE SPACECRAFT

Following the excessive waste of the throwaway Apollo era, one of the primary factors for the engineers designing the Space Shuttle was to come up with a launch system that was reusable. In its final form the only major component of the STS not refurbished is the ET which is jettisoned by the Orbiter to burn up in the atmosphere. The remaining parts – the two SRBs and the Orbiter itself – are reused time and time again. Out of the five spaceworthy Orbiters, *Discovery* holds the record with thirty-seven flights (as of October 2009) and two more are scheduled for 2010.

RECOVERING AND REFURBISHING THE SRBs

Both SRBs separate from the Orbiter 124 seconds after launch at a height of around

◀ *Lift-off of* Discovery *on mission STS-64.*

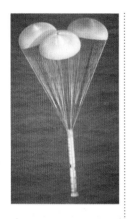

A Solid Rocket Booster (SRB) falls back into the Atlantic ocean with all three chutes open.

▶ *The large External Tank (ET) shown shortly after separation. This is the one component that is not reusable and burns up during re-entry through the atmosphere.*

16,000ft (48,800m) and they continue upwards to their highest point, apogee, at 220,000ft (67,000m). By this time the nose cones have been ejected and first a small pilot parachute and then a 54ft (16m) drogue is deployed to reorientate the SRB tail-first for main parachute activation. There are three of these massive chutes, the largest ever used. Each has a diameter of 136ft (41m) and a design load of approximately 195,000lb (88,000kg). About 279 seconds after separation, the SRBs splashdown in the Atlantic, 150 miles (240km) off the eastern coast of Florida.

NASA has two recovery vessels, *Freedom Star* and *Liberty Star*, which locate the SRB casings and the chutes, including pilot and drogue which are also recovered. Divers place plugs into the rocket nozzles to seal the SRBs and after pumping out they are

'And as we know now, and as I pointed out many times, the great plume of fire at the bottom of the Space Shuttle is actually dollar bills burning, and the most efficient method of destroying American dollar bills as has ever been devised by man.'

Dana Rohrabacher, Chairman of the Subcommittee on Space and Aeronautics – during fiscal year 1998 NASA authorization hearings, March 1997.

The NASA recovery ship Freedom Star *with a spent SRB from STS-114 in tow, passing through Port Canaveral.*

Did you know?
The Orbiter's undercarriage is manually activated by the crew. And because the three wheels are lowered through flaps in the heat shield they cannot be retracted once deployed.

towed in a horizontal position back to Cape Canaveral and KSC. Once they are taken out of the water they are loaded on to modified rail wagons and moved to a hangar where they are given a thorough post-flight inspection before the individual segments are separated. The sections are sent to Morton-Thiokol in Utah where they are cleaned, inspected and then refilled with the solid fuel; an ammonium perchlorate composite propellent. This whole process can take anywhere from six to twelve months before the refurbished segments are returned to KSC ready to be used again.

After landing at Edwards AFB, California, Atlantis is being mounted atop the special Boeing 747 Shuttle Carrier Aircraft (SCA) for the ferry flight back to KSC.

REFURBISHING SPACE SHUTTLE ORBITER

Even though most landings take place at the Shuttle Landing Facility (SLF) near KSC, there are still occasions when adverse weather conditions in Florida dictate a landing at Edwards AFB, California. To ferry Orbiters back to KSC NASA has two special Boeing 747 Shuttle Carrier Aircraft (SCA). The first of these, originally manufactured for American Airlines, was used to carry and release *Endeavour* OV-101 during the ALT programme in the 1970s. The newly returned spacecraft is hoisted into position for its piggyback ride using the Mate-Demate gantry located at the Dryden Flight Research Centre at Edwards. A similar gantry lifts the Orbiter off the 747 after its arrival at KSC.

All returning Orbiters are sent to the Orbiter Processing Facility (OPF), a group of three hangars located near the Vehicle Assembly Building (VAB). Inside the OPF the vehicle is made safe with all residual propellants drained, and any returning payloads are unloaded. The Orbiter's systems and equipment are extensively inspected, and refurbished or replaced as appropriate. For example, the SSMEs are removed and go back to the engine shop for detailed examination. Normally a new set is ready to be installed.

As is to be expected, the Thermal Protection System (TPS) comes under particular scrutiny and any damaged tiles are carefully removed and a replacement bonded into place. Earlier Orbiters had around 34,000 tiles while nowadays they have about 23,000 as less vulnerable areas are now protected with thermal blankets to reduce maintenance and weight.

Any large horizontal loads, such as the Spacelab modules, are installed in the payload bay, and once all systems tests and checks have been completed the Orbiter is transferred to the nearby VAB. (Vertical payloads will be loaded when the Orbiter is in the upright position.) It used to be the practice to tow an Orbiter from the OPF to the VAB on its own undercarriage, but the

> 'The thing I'll remember most about the flight is that it was fun. In fact, I'm sure it was the most fun that I'll ever have in my life.'
>
> Sally K. Ride, first woman astronaut to fly in the Space Shuttle, 1983.

▼ Endeavour *riding piggyback on the Boeing 747 SCA on its way back to KSC, December 2008.*

▶ Discovery *inside the Orbiter Processing Facility (OPF) at KSC, marking the completion of mission STS-114 in 2005.*

▶▶ Columbia *being returned to the Vehicle Assembly Building (VAB) in preparation for its return to space on mission STS-93 in 1999.* Columbia *was the first of the Orbiter fleet to fly, in 1981, and made a total of twenty-eight flights into orbit before its destruction during re-entry in February 2003.*

engineers found that they were picking up debris on the tyres, so now it is transported on a special flatbed trailer in an operation still known as 'roll-over'.

Inside the VAB the Orbiter is hoisted into the vertical position and mated with the ET and SRBs, and once a positive Flight Readiness Review has been completed the world's only reusable launch system is ready to go into space again.

END OF AN ERA

One of the main recommendations to emerge from the inquiry into the *Columbia* accident was that the Space Shuttle should not fly beyond 2010 unless the fleet was fully re-certified. Accordingly, President George W. Bush ordered their retirement after almost thirty years in service. *Atlantis* will be the first of the three remaining Orbiters to be withdrawn, followed by *Endeavour* and then *Discovery* with the final mission, STS-134, scheduled for September 2010.

Replacing the Shuttle will be new generation of boosters and spacecraft within the Constellation programme which builds upon many of the lessons learned with the Shuttle. Constellation consists of two main elements with a new launch system, or systems, and a Crew Module (CM) which resembles an Apollo Command Module on steroids. Unlike Apollo, the Orion CM will be reusable for up to ten flights and will have the capacity for up to six crew members. Initially it is planned for low-orbit missions such as rendezvous and docking with the ISS, but future variants may take America back to the Moon in conjunction with the proposed Altair Lunar Lander, perhaps by 2020.

Two launch systems will be developed as part of Constellation. The smaller

Did you know?
Former Apollo 11 astronaut Buzz Aldrin is calling for an international collaborative effort as the only way for mankind to return to the Moon.

◄ *Shuttle for sale? Dale Gardner held up the 'For Sale' sign after the* Discovery *STS-51-A crew had retrieved the Palapa B-2 and Westar 6 satellites after their payload assist modules failed to fire.*

▶ A night launch of STS-119 in March 2009, taking Discovery into orbit on an ISS truss and solar array assembly flight.

▶▶ The Lockheed Martin X-33 Venture Star might have been the shape of things to come if this proposed design from the 1990s for an unmanned single-stage-to-orbit reusable launch vehicle had gone ahead.

▶▶ Artist's concept image of the heavy-lifting Ares V during SRB separation. Ares V is the cargo launch component of the upcoming Constellation programme to replace the Shuttle.

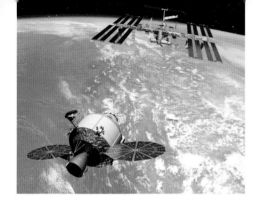

> 'Some things simply are inherent to the design of the bird and cannot be made better without going and getting a new generation of spacecraft. That's as true for the Space Shuttle as it is for your toaster oven.'
>
> Michael Griffin, NASA Administrator – on Space Shuttle safety, eve of launch of STS-114, July 2005.

◄ Looking like a beefed-up Apollo spacecraft, the Orion is depicted here approaching the ISS.

Ares I is designed to place the Orion CM into orbit and will feature a single SRB beneath a liquid-fuelled second stage. Its big brother is the Ares V, a massive booster more powerful than the Saturn V, capable of delivering a payload estimated at 414,000lb (17,800kg) into low-earth orbit or 157,000lb (71,200kg) to the Moon. Ares V will incorporate five main engines assisted by two SRBs, plus an upper Earth Departure Stage (EDS) which can either

◄ The end of the Shuttle. Discovery STS-120 touching down at KSC. In September 2010 Discovery is scheduled to make the final flight of the Shuttle programme.

send the Orion/Altair combination to the Moon or deliver large modules to the ISS.

The Constellation programme is intended to provide a flexible launch system which can be adapted to a number of roles, but it is not without its detractors. Concerned about the cost, in 2009 President Obama ordered a review of the programme which indicated that manned flights to the Moon or Mars were beyond the current budget, and there is an ongoing debate regarding whether both the Ares I and Ares V launchers are needed.

The immediate dilemma now facing NASA is how to fill the void between the Shuttle's retirement and Constellation's entry into service. Unwilling to extend the Shuttle's service life they will become dependent upon Russian spacecraft in the interim, and a report published in *New Scientist* in 2007 suggested that NASA planned to seek an exemption to a congressional ban that prohibits the purchase of Russian *Soyuz* rockets. Not surprisingly there have been some calls in the US Congress to keep the Shuttle operating until Orion/Constellation is up and running in 2014 or 2015. However, NASA's bosses believe that any extension of the Shuttle programme past 2010 would be unsafe, counter-productive to the development of the new spacecraft and very expensive. In truth the Space Shuttle has reached the end of its useful life.

'The more of us beyond the Earth, the greater the diversity of worlds we inhabit ... then the safer the human species will be.'

Carl Sagan.

APPENDIX 1 – SEEING THE SPACE SHUTTLE

If you are quick you might just catch one of the last few launches before the Space Shuttle's retirement which is scheduled for September 2010. NASA has offered *Discovery* for display at the Smithsonian National Air & Space Museum as part of the national collection, and suitable homes are also being sought for *Atlantis* and *Endeavour* once they are decommissioned. In the meanwhile *Enterprise* and a number of non-flying mock-ups are already on display:

SPACE SHUTTLE *ENTERPRISE* OV-101

The first Space Shuttle, *Enterprise* never did follow its TV namesake into space (see 'The Orbiter Fleet'). After five free flights to undertake approach and landing tests it went on an international promotional tour to Europe and Canada riding piggy-back on the SCA before being handed over to the Smithsonian Institution in November 1985. It is now on show at the NASM's Steven F. Udvar-Hazy Center at Dulles International Airport, Washington DC.

www.nasm.si.edu/museum/udvarhazy

SPACE SHUTTLE *PATHFINDER* OV-098

Built in 1977, this simulator was used to test ground handling facilities, such as roadway clearances, and for ground crew training at KSC. Sent for display in Tokyo, from 1983–1984, it has returned to the

◀ Pathfinder *mock-up at the US Space and Rocket Center in Huntsville. (Michael Fallows)*

➤ Enterprise *on display in the NASM collection at Dulles. (Ad Meskins)*

USA and now named *Pathfinder*, with the honorary OV-098 designation, it is currently on display at the US Space and Rocket Center in Huntsville, Alabama.

www.spacecamp.com/museum

SPACE SHUTTLE *EXPLORER*

Explorer is a scale replica, including interiors, at the Kennedy Space Center visitors' complex, Florida.

www.nasa.gov/centers/kennedy

SPACE SHUTTLE *ADVENTURE*

Full-scale mock-up of an orbiter mid-deck and flight deck at the Johnson Space Center in Houston, Texas.

www.spacecentre.org

SPACE SHUTTLE *AMERICA*

Not constructed by NASA, this full-scale mock-up is situated at the Six Flags Great America theme park in Gurnee, Illinois.

SOVIET *BURAN* SPACECRAFT

Although the original unmanned *Buran* was destroyed when the roof of the building where it was being stored collapsed in 2002, a number of test vehicles and mock-ups have survived. Those on display include the BST-002 jet-engined flying test vehicle at the *Technikmuseum Speyer* in Germany, and a non-flying structural test vehicle at Gorky Park, Moscow.

www.technik-museum.de

APPENDIX 2 – TECHNICAL SPECIFICATIONS

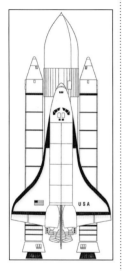

The specifications and performance of the Space Shuttle have changed during its period of operation, largely due to weight reductions and other improvements. These figures are based on the fifth and final model, *Endeavour* OV-105.

SPACE SHUTTLE LAUNCH SYSTEM:

Overall height	184ft (56.1m)
Gross lift-off weight	4,400,000lb (2,000,000kg)
External tank length	154ft (46.9m)
External tank diameter	28ft (8.4m)
External tank propellant volume	534,900USgal (2,025cu/m)
SRB length	150ft (45.6m)
SRB diameter	12ft (3.7m)
SRB thrust at lift-off	12.5 million newtons

SPACE SHUTTLE ORBITER:

Crew	5–7 normally, can range from 2–8
Length	122.17ft (37.24m)

Wingspan	78.06ft (23.8m)
Height on runway	58.58ft (17.86m)
Empty weight	172,000lb (78,000kg)
Gross lift-off weight	240,000lb (110,000kg)
Max landing weight	230,000lb (100,000kg)
SSME main engines	3 x Rocketdyne Block IIAs
Speed	17,320mph (27,870km/h)
Max payload	55,250lb (25,060kg)
Payload bay length	59ft (18m)
Payload bay width	15ft (4.6m)

MISSIONS

Duration	From 5 to 16 days, the longest mission was 17.5 days on STS-80 in 1996
Orbit height	Between 115 to 400 miles (185–643km)

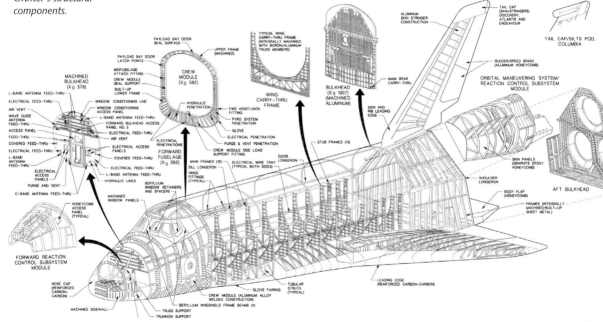

Diagram of the Orbiter's structural components.

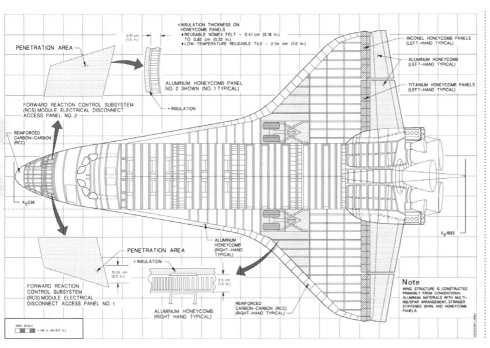

Plan view of the Orbiter's structure.

> NASA drawing showing main component system location.

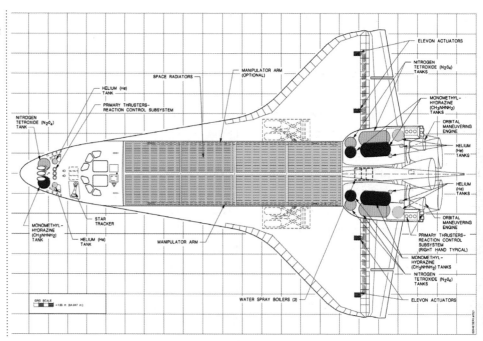

APPENDIX 3 – GLOSSARY

ALT	Approach and Landing Test programme	IUS	Inertial Upper Stage – launches payloads into higher orbit
AOA	Abort Once Around	KSC	Kennedy Space Centre
APU	Auxiliary Power Unit	MLP	Mobile Launcher Platform
ATO	Abort To Orbit	MMU	Manned Manoeuvring Unit
DoD	Department of Defense (US)	MPLM	Multi-Purpose Logistics Module
DPS	Data Processing System	NASA	National Aeronautics & Space Administration
ESA	European Space Agency		
ET	External Tank	OMS	Orbital Manoeuvring System
EVA	Extra-vehicular Activity	OPF	Orbiter Processing Facility
FSS	Fixed Service Structure	OVD	Orbital Vehicle Designation
GLS	Ground Launch Sequencer	PAM	Payload Assist Module
HRSI	High-temperature Reusable Surface Insulation	RMS	Remote Manipulator System – robot arm
HST	Hubble Space Telescope	RTLS	Return To Launch-Site – launch abort mode
ISS	International Space Station		

SAFER	Simplified Aid For EVA Rescue – propulsive backpack system	SSME	Space Shuttle Main Engine
SAS	Space Adaptation Syndrome – space sickness	STS	Space Transportation System – NASA's designation for the Space Shuttle
SCA	Shuttle Carrier Aircraft	TAL	Transoceanic Abort Landing
SLF	Shuttle Landing Facility	TPS	Thermal Protection System
SRB	Solid Rocket Booster	VAB	Vehicle Assembly Building